Careers

Production

Careers

in
Film
and
Video
Production

Michael Horwin

Focal Press
Boston   London

Library of Congress Cataloging-in-Publication Data

Horwin, Michael.
    Careers in film and video production / Michael Horwin.
        p.   cm.
    Bibliography: p.
    Includes index.
    ISBN 0–240–80049–4
        1. Motion pictures—Production and direction—Vocational
    guidance.   2. Video recordings—Production and direction—
    Vocational guidance.
    I. Title.
    PN1995.9.P7H67   1989
    791.43'0232'023—dc20   89–34330
                                                            CIP

British Library Cataloguing in Publication Data
Horwin, Michael
    Careers in film and video production.
    1. Cinema films. Production
    I. Title
    791.43'0232

    ISBN 0–240–80049–4

Butterworth–Heinemann
80 Montvale Avenue
Stoneham, MA 02180

10   9   8   7   6   5   4   3   2

Printed in the United States of America

To

Raphaële, whose loving support made this book possible.

Amie, Diane, and Louis Horwin—my best fans.

And to the living memory of Ray Horwin.

Contents

Part IV: Producing and Directing

Contents _____ix

Preface

The objective of *Careers in Film and Video Production* is to explain how film and video production really works to those individuals who wish to understand it better and are considering pursuing a career in it. The book is written as a practical manual to give students of film and video a place to begin in exploring the variety of positions in production and in understanding the steps required to pursue those jobs.

The motion picture/television industry is one of the most powerful tools in modern society. The style and content of movies, television, commercials, videos, documentaries, and industrial and educational films shape the way people perceive the world. Those who control the media and decide what is produced and shown to audiences and what is not have the power to manipulate public opinion, attitudes, and priorities.

In 1791, the framers of the Constitution added the Bill of Rights, in which the first amendment stated, "Congress shall make no law . . . prohibiting" people from having "freedom of the press." This proclamation did not guarantee that each individual would have the opportunity to use the press, only that no law would be passed to prohibit such use. Today, some people argue that the only individuals who have freedom of the press (the media) are those who own the media or are paid to work within it.

Perhaps this is one reason why the motion picture/television industry has, for many years, had an aura of elitism. Each job was hotly contested, and those with relatives in the industry had the advantage. This is less true today. As productions continue to increase in number and geographically disseminate from tiny centers of production, there is a greater opportunity to get started. Like any powerful industry that pays well, however, this industry will never have complete democratization. It takes at least one million dollars to produce a feature-length film; this automatically leaves out many of us as financiers (i.e., executive producers). However, the best way to gain access to the opportunity to make your own films is to contribute to the production of someone else's film and, in the process, to learn what filmmaking is about from the inside. It is unrealistic to aspire to start as a director or a producer, but it is completely realistic to get into film at some level, learn more about production as you work, and then move up the ladder as quickly as your talent, abilities, and opportunities allow.

One producer explained, "I've worked my way up the ranks. I've done it all. If you're going to be good it takes time, but whenever the opportunity arises and you have the chance to do what you really want to do—then you'll be ready for it."

The first step to making inroads into the motion picture/television industry is understanding how films are made, the subtle interaction of business and art, the various jobs in production, the language spoken on the set, and how to start working. *Careers in Film and Video Production* was written to provide you with that information. It discusses what happens behind the scenes and behind the camera; the talents, abilities, and determination a person needs to succeed in getting into the motion picture/television industry; and what the director and producer do during preproduction, photography, and postproduction, the three stages in producing a film or video. It also discusses the jobs and responsibilities of the "unsung heroes," the technicians such as the grips, electricians, and gaffers; the makeup artists and hairstylists; the production accountants; the transportation coordinators; the location managers; the production assistants; the production designers; the art directors; the camera operators; and others.

Part I begins with what it means to work in production—the knowledge required and the life-style. Part II explores what production is and how films and videos are actually produced. Part III discusses eight fundamental job categories and six support services. Each chapter in this section provides a career profile, a discussion of duties and responsibilities, verbatim interviews with professionals in each area, and pragmatic suggestions on how to begin working. Part IV examines the business of producing and the art of directing through an interview with an eminent producer (Ronald G. Smith) and director (Oliver Stone). And, finally, a glossary of film and video terminology and a variety of appendixes that list sources of production information are provided at the end of the book.

Careers in Film and Video Production was not written by researching other books. There are no other books on this subject. The research methods consisted of interviews with some of the top people working in film and video production today. Questions included the following: What is your job? How did you learn your job? How did you begin working?

Because most of the actual filmmaking takes place during the production stage (there are three stages: preproduction, production, and postproduction), *Careers in Film and Video Production* focuses more on production than on any other phase. In addition, production, or shooting, involves the greatest number of jobs and the best opportunity to examine the nuts and bolts of filmmaking. The specific positions discussed in detail in Part III have been chosen because they represent many of the most critical positions that most productions cannot do without. A second reason why these particular jobs are discussed is because they provide some of the best opportunities for newcomers to get started.

To summarize, *Careers in Film and Video Production* is designed to do two things: (1) discuss the opportunities for people with dedication and talent to become part of this exciting industry, and (2) instill a greater appreciation for how films and videos are produced by examining the creative and technical processes involved in production. Filmmaking is a conjurer's art, and by opening the door on a closed set and glimpsing the people, the jobs, and the tools that create the magic, it is possible to have a greater understanding of the medium. Film, at its very best, represents the ultimate art form. It is the only medium that encloses and translates all others—dance, music, painting, and literature. The way in which film is created affects our appreciation of the other arts and our perception of the world. Those who take part in creating film are able to communicate their thoughts, artistic sensibilities, and values to great numbers of people and to leave their mark behind.

Acknowledgments

To thank all the people who contributed to this book, I would have to list everybody who taught me about film and video production. I stand in particular debt to Ronald G. Smith, who went out of his way to listen to me plead for my first job, and to Andrew Linsky, who, whether he knew it or not, taught me about production.

I am especially grateful to all my interviewees, the working professionals who took time from their own responsibilities to answer many questions. These include Dan Adams, Michelle Buhler, Jane Galli, Susan Gelb, John Grant, Bruce Greene, George Herthel, Willy Mann, Alan McKay, Michelle Minch, Dan Morski, Mike Paris, Rick Rollison, Mike Sheperd, Howard Smit, Ronald G. Smith, Paul Steinke, and Robert Studenny.

I also wish to express my sincere gratitude to Mr. Oliver Stone, a softspoken man of power and thought. Mr. Stone and his films serve as a reminder of the greatness of this medium and the kind of work to which everyone beginning in this industry can aspire. I thank Mr. Stone for taking time from his production to speak with me about the art of directing.

Part 1

Working in Film and
Video Production

Getting into Production

The production of movies or feature-length films was once the sole possession of a few tightly controlled major studios. Today, the industry has changed. With the advent of the ancillary markets (cable television and videocassette) and the continued growth of foreign sales, the production of films by independent filmmakers has surpassed the numbers produced by the majors. In fact, the large studios have used this trend to their advantage by hiring independent producers to create relatively low-cost films that they distribute.

The birth of the ancillary markets, the growth of independent production, and the decentralization of the motion picture/television industry have all contributed to create a boon of opportunity, jobs, and challenges to a young and growing film industry. There are more regions where films are being produced and a greater number of productions each year.

The term *motion picture/television production* includes the making of movies or feature-length films, television shows, television commercials, educational and industrial films, documentaries, and videos. For the purposes of this book, *film* production or *filming* can mean the production of film or video. (Because the video industry is an outgrowth of the film industry, film terms are often used interchangeably with the more precise video terminology.)

What Is Required

Working in film requires a rare sort of dedication. It requires persistence and confidence, and it requires information. You must understand how films and videos are produced, what the roles and responsibilities are of the people on the production team, and how the various positions work together to create a finished product.

The production team may be comprised of a hundred or more skilled professionals working in distinct and separate areas, from the director and producer at the top to the P.A. or runner at the bottom and everyone in between including: the location manager, hair and makeup artist, production accountant, grip, electrician, driver, transportation captain, art director, set dresser, camera assistant, and many other departments and individuals.

It is important to be aware of many of these positions. After all, how can you get hired for a position that you do not know exists, doing something that you have never heard of?

Getting Started

There is an often-repeated remark that breaking into the film industry is impossible. It is true that you will not find production jobs listed in the want ads of your classified newspaper, and employment agencies usually do not list new openings in film. Because these traditional methods of getting a job do not work in the motion picture/television industry, this industry has won the reputation, justly deserved or not, of being a closed shop.

It is not closed. In fact, many of the most skilled artists, technicians, and craftspeople working in production today had other careers five, ten, fifteen, and twenty years ago. There is a constant flow of new people into this industry as the industry continues to expand.

To say that there is a formula to breaking into film is to disregard the whole creative process for forging a career in the motion picture/television industry. For nearly every person working in this business there is a different way someone got started. There is no royal road unless you are independently wealthy. However, there is a commonsense approach that has long been successful and that takes into consideration the film industry's unique qualities.

The motion picture/television industry is different from other industries in many respects, including the self-training that is required and the large role of a freelance work pool.

Importance of Self-Training and Freelancing

Self-training is the single most important aspect to working in production. To prepare yourself to be a banker or real estate agent, you can enroll in a class and be trained while on the job. But, except for film schools and universities, very few organized training environments prepare people for work in film production. Many people who have earned paying jobs in the motion picture/television industry, both film school and non-film school graduates, trained and prepared themselves.

The second quality important to starting in production is understanding the freelancer's role and responsibilities. Most productions last for months or weeks, depending on the film's finished length. Therefore, the inherent nature of production is to hire personnel for similarly intermittent short periods. Most people in film work freelance and are highly motivated self-starters who are employed on a per-job basis, sometimes day to day, often week to week, and occasionally month to month.

Therefore, to take advantage of opportunities that exist in the motion picture/television industry, it is important to:

1. Self-train, which means knowing the vocabulary of production, how a film or video is produced, and the roles and responsibilities of the various departments in film.

2. Know how to work as a freelancer, which includes understanding film etiquette, how to use the telephone, how to network, and how to determine competitive rates for various types of work.

These are the fundamental ingredients in the recipe to get hired and to stay working.

Another reason this approach works is because a film company generally cannot afford to provide on-the-job training, since a huge amount of money is at risk in a production. Production in and of itself is a risky business. The purpose of a film or video is to entertain (e.g., movie or television show), sell (e.g., commercial) or educate (e.g., educational program, documentary). The success of 1000 feet of film or videotape can only be measured after it is produced. Other products can rely on prototypes, but a film or video must be completely created, with a great deal of money expended, before it can be judged as a complete success or failure. Did the commercial pull? Did the music video increase record sales? Was the movie successful at the box office?

Financially it is an expensive risk. Hundreds of thousands of dollars are spent to produce thirty seconds of commercial. The day rate of an experienced director or star may be over $3000, and it may cost $10 to $20 million to produce a feature film. The numbers are astounding, and still the medium itself is inherently risky. (Watch the disappointing volume of trade on the New York Stock Exchange for the major motion picture studios.) Since the production's ultimate fiscal success takes the form of a large question mark that looms after most of the money has been spent, the crews and artists are expected to be known commodities, constants in a world of variables. As constants, each member of the production team is expected to successfully fulfill his or her responsibilities. There is simply too much at risk in the long run to have problems in the short run. Therefore, production companies depend on a skilled labor pool of experienced film and video professionals to get the job done. Newcomers are expected to bring some knowledge of production to their first day on the job.

Any experienced crew member will tell you that an error on the set only becomes a mistake when it costs the production time or money. For this reason, there is an inherent weeding-out process in the film industry. Individuals must pay their dues and work on whatever low-budget shows they can to get trained before they get a chance to work on productions with larger budgets. That is why self-training is important, and that is why you will seldom find nonexperienced people working on any sizable production.

The Vocabulary of Production

The sunset is beautiful, and the director decides he wants to roll some film. The production manager sends a P.A. on an emergency run to find Eric the D.P. The P.A. runs to base camp. In a few minutes he returns with Eric the propmaster. The sun is sitting on the horizon. That is a problem.

A new location scout finds a great house, and the director loves the pictures. The production manager pulls the scout aside and says, "Remember there's prep and wrap." The scout negotiates for the one-day shoot. The day before filming commences, the set decorator wants to get into the house. Nobody is home. That is a problem.

Every industry has its own language, and the film industry is no exception. What separates the language of the film industry (or any industry) from mainstream colloquial American language is its use of specialized names for certain jobs, tools, and events.

Special languages allow members of the same industry to communicate effectively, and in a film and video vocabulary, words may have a different meaning in "filmspeak" than they do in mainstream conversation. Some examples include rolling, speed, fishpole, sticks, flag, and grip. In filmspeak, *rolling* means that a recording device has been switched on. *Speed* means that the tape within the device is rolling at sufficient speed to record correctly. *Fishpole* is a kind of extension rod on which a microphone is attached. *Sticks* is another name for tripod. *Flag* means a sheet of black material used to create a shadow. *Grip* is a job designation used to describe those individuals who move the camera and create the shadows. An individual in the film industry should be familiar with some of these examples in case she or he is asked to "tell the grip to bring a flag, then grab the sticks and find the boom operator with the fishpole."

The more specialized an industry is, the more specialized is its vocabulary. The more innovative an industry, the more rapidly new words evolve. The film industry is both specialized and innovative, and therefore it has a large vocabulary with new words added each year. *Shotmaker* and *steadi-cam* are words for new tools that have recently been invented.

But an industry-based language does more than allow individuals to communicate effectively with each other. It also allows individuals to differentiate quickly who is in their industry and who is not. Language is one of the badges that individuals wear to signal their identity with a specific group.

For both these reasons, effective communication and identity, it is important to be familiar with filmspeak. You must wear the badge so that people will feel comfortable hiring you. You will signal to them that you belong. And once on the set you will know what D.P. means if you are told to find him or what prep and wrap means if you are responsible for negotiating prep and wrap time at a location.

You should become familiar with the most popular terms for various jobs, tools and events. (A glossary of over 200 filmspeak words is found at the end of the book.)

The Jobs

A helpful hint in defining the various jobs is to pay more attention to the second word in any two-word title. The first-word qualifier in these titles can become awkward in the beginning, as in production assistant, production coordinator, production manager, production accountant, production designer, set decorator, and set dresser. If you remove the first-word qualifiers you have assistant, coordinator, manager, accountant, designer, decorator, and dresser. Although it is rather simplistic, by changing the remaining titles into verbs you can find out exactly what they do on a production. The assistant assists, the coordinator coordinates, the manager manages, and so on. Another way to grasp what the jobs are is to simply reverse the two words in the title. The production manager manages the production, the production assistant assists the production, the production designer designs the production, and so on. Most titles that begin with a word other than "production" signify a subdepartment within the production itself. The production manager manages the entire production, the location manager manages only the location department within the production, the transportation coordinator coordinates the transportation department, and so on.

Another hint is that any title with the word "key" signals a department head. The key grip is in charge of the grip department. Key props signifies the head of the prop department.

Of course, knowing that someone else "manages" while another person "coordinates" does not explain what they do. At this point it is helpful to meet some of the various positions and departments involved in the production of a film or video. The following list does not contain every job in production, but it does include many of the departments that are part of most productions most of the time during a shoot.

PRODUCER

Usually the first person on a project, the producer develops, finances, and oversees the entire production from beginning to end.

DIRECTOR

The director is responsible for all creative aspects of the production. It is the director's vision that will be translated onto film or tape.

First Assistant Director

The first assistant director (first A.D., and also called the "first") is the liaison between the director and the production manager. The first A.D. is in charge of organizing the shooting schedule and making sure that the crew, cast, and equipment needed on the set are there when they are supposed to be. The first A.D. also maintains order on the set and may direct extras and crowd scenes.

Second Assistant Director

The second assistant director is the first A.D.'s right hand. The second handles much of the paperwork, including preparing and distributing call sheets and production reports, and may assist the first in maintaining order on the set.

SCRIPT SUPERVISOR

The script supervisor takes detailed notes of each scene and take, including camera position, dialogue changes, and the running time of each shot.

LINE PRODUCER

This production-oriented producer supervises the entire production. (The line producer may be the same as the producer.)

PRODUCTION MANAGER

The production manager, also called the unit production manager (U.P.M.), is responsible for overseeing all financial and administrative aspects of the production, for budgeting expenditures, for hiring the various department heads, and for overseeing the entire crew.

PRODUCTION COORDINATOR

The production coordinator (also called production office coordinator) works for the production manager. Responsibilities include acting as liaison between the production manager and other crew members for much of the production's clerical work.

DIRECTOR OF PHOTOGRAPHY

The director of photography (D.P.), also called the cinematographer (or cameraman) is responsible for the photographic images of the production, including the lighting, framing, and exposure of each shot.

CAMERA OPERATOR

The camera operator is responsible for actually running the camera and keeping the action in frame.

First Assistant Cameraman

The first assistant cameraman assembles, cleans, and maintains the camera, sets the focal aperture, measures the distance between the object being photographed and the lens, keeps the camera in focus when the person or object is moving toward or away from the camera, and places marks (where actors will stand).

Second Assistant Cameraman

Second assistant cameraman or loader, loads and unloads the magazines (film container), claps the slate, and writes the camera reports, which list the scenes and takes for each shot and detail what should be done with it (i.e., print).

GAFFER

The gaffer is the chief electrician, the head of the electrical department, and is responsible for execution of the lighting as described by the D.P.

Best Boy Electrician

The best boy electrician, or first assistant electrician, is in charge of the equipment and administration of the electrical department.

Electricians

The electricians comprise the gaffer's crew and are responsible for powering the lighting and electrical equipment and positioning the lights.

KEY GRIP

The key grip is the head of the grip department and is in charge of camera movement, creating shadows, and rigging lights and camera.

Best Boy Grip

The best boy grip oversees the equipment and administration of the grip department.

Dolly Grip

The dolly grip operates and maintains the dolly and crane equipment.

Grips

The grips are the workers in the grip department.

PRODUCTION MIXER

The production mixer or sound recordist operates the sound recording equipment. During takes that use two or more microphones, the mixer balances the levels and equalization or "mixes" the different signals. The sound mixer also monitors the recording and keeps detailed sound reports.

Boom Operator

The boom operator handles the microphone boom, a long pole that allows the microphone to follow the action.

PRODUCTION DESIGNER

The production designer is responsible for establishing the "look" of a picture and supervising the production's visual design.

ART DIRECTOR

The art director is responsible for designing the sets.

SET DECORATOR AND SET DRESSER

The set decorator is the interior decorator of the set, and the set dresser assists the decorator during the shoot.

PROPERTY MASTER

The property master is responsible for obtaining and placing all props on a set.

COSTUME DESIGNER

The costume designer designs and supervises the creation of all costumes in accordance with the design established by the director (and possibly the production designer).

COSTUMER

The costumer is responsible for maintaining the costumes during production.

MAKEUP ARTISTS AND HAIRSTYLISTS

The key makeup artist is responsible for the actor's makeup (generally makeup applied to the head, face, hands, and lower arms) and may oversee the job of the body makeup artist (who applies makeup from the neck down), hairstylist (who styles the actor's hair, wigs, and toupees) and other assistant makeup artists.

SPECIAL EFFECTS

The special effects department may comprise a technician or an entire company and is responsible for the planning and safe execution of all special effects, which may include everything from a working dishwasher to floods, fire, explosions, and special visual effects.

STUNT COORDINATOR

The stunt coordinator is responsible for planning and safely executing all stunts on a production.

LOCATION MANAGER

The location manager is the head of the location department and is responsible for scouting locations, negotiating the rental of the locations, securing all government permits, and organizing location details including parking, catering, and fire and police safety officers.

STILL PHOTOGRAPHER

The still photographer is responsible for taking still photographs of the production. The photographs may be used to help match continuity in later scenes or for publicity purposes.

CONSTRUCTION DEPARTMENT

The construction department is composed of carpenters, painters, plasterers, welders, paperhangers, prop makers, and other laborers and is headed by a set construction foreman. This department is responsible for constructing all sets in and out of the studio.

TRANSPORTATION DEPARTMENT

The transportation department, under the transportation coordinator, is responsible for everything that rolls in a production, including all equipment vehicles used behind the scenes and all picture vehicles used in front of the camera.

PRODUCTION ACCOUNTANT

The production accountant (production auditor or controller) oversees the accounting department, which handles the production's bookkeeping and estimating.

PRODUCTION ASSISTANT

Production assistants (P.A.s or runners) are "catch all" assistants who support most of the other departments in production.

The Tools

The motion picture/television industry relies on thousands of special tools to create its illusions. Every department in a production has its own tools. From menthol blowers to production reports, from dolly track to halogen medium iodide lamps and camera cars, it is easier to learn what each of the departments does than to remember the names of the hundreds of tools that each department relies on.

However, learning about the tools is one of the most fascinating aspects of learning about film. Understanding mattes and foleys is like watching a sleight-of-hand trick in slow motion. Appreciating how a dolly track is laid down and how the focus is racked provides a new perspective and appreciation in watching any scene.

It is important to understand that the film industry uses many tools specially created for its own use and that the names and functions of these tools may apply only to their operation in film. This is especially true for the technical departments concerned with camera, lighting, and postproduction. Cinematographers, camera operators, grips, electricians, and editors rely on highly specialized equipment created specifically for them. Therefore, many of the terms used in these areas have no equivalents or near equivalents in common speech.

Although you cannot learn the names of all the tools immediately it is helpful to be aware of some of the names that occur most frequently. The miniglossary on the next page has been created to familiarize you with eleven of some of the most basic tools. (For a more complete list of tools, see the glossary at the end of the book.)

Another "tool" frequently bandied about in production is the application of the "line." The production manager divides the production's budget into "below the line" and "above the line" expenditures. Generally, "below the line" expenditures include the fixed costs of hiring the crew, renting equipment and materials, location costs, and film stock. "Above the line" expenses are usually the most expensive items and include the costs of the producer, director, principal cast, and screenplay.

Tool	Description
Key light	Main source of light in a scene. Provided by electrical department.
Sandbag	Burlap or plastic bag filled with sand used to hold down pieces of equipment.
Camera car	A special vehicle that carries the director, D.P., and other members of the crew when filming a moving vehicle or person.
Changing bag	A lightproof bag used by the film loader to load film into the magazine.
Magazine	Film container or cassette that fits onto the camera.
Prop	Any "handable" or portable item used in a scene or that is specifically mentioned in the script.
Scrim	Translucent material placed in front of a light source to diffuse the light.
Slate	Hinged board (clapper) used in the beginning or end of a take to provide a cue for synchronization of sound and picture and to provide information about the scene and take (e.g., scene number, take number, roll number).
Apple box	A wood crate used to raise the height of props, lights, and other objects and people.
Barn doors	Folding metal gates in front of a lamp (light) that control the direction and amount of light.
Marks	Masking tape or chalk put on the floor and used to indicate where camera and actors should stop during a movement.

The Events

In filmspeak the major events are (1) preproduction, where everything that must be prepared for photography takes place; (2) photography, also called principal photography, production, or shooting; and (3) postproduction, where the film or videotape is edited, music is added, and so on.

The set is also an event in the sense that it is what becomes of a location (a stage or a "practical" location such as a real house, street, or building) immediately before, during, and after the cameras are brought in and filming takes place. The term used to describe the preparation of a set to be filmed or taped is *prep*. During prep, the location is dressed (made to look as it should) and perhaps pre-rigged (rigged with lights to save the crew time during the day or night of the shoot). After filming or taping is completed at the location, the set is "struck," or torn down, and the location is returned to its original condition. Prep and strike can occur on the same day as the shoot or, depending on the complexity of the set, can take large crews working many days before and after filming. Some sets are prepped weeks before shooting begins and require several days for strike. Other events include "takes" and "scenes." A take is a segment of film or videotape shot at a single time; a scene is one or more shots that present a main action at the same location. The glossary at the end of the book lists many of the other events in filmspeak.

When the following paragraph makes sense to you, you will know that you are beginning to understand filmspeak.

"Listen, we're losing daylight and we have to get the shot off. Go to the grip truck, it is parked next to the jenny, and bring an applebox, a c-stand, and a sand bag down to the set. Then go over to the honeywagon and wait for the steadi-cam guy. When you find him, tell the first A.D. Later we may need you to prep the picture cars and help the grips rig a butterfly. When we're wrapped, get the other P.A. to help strike this place. And by the way, you are doing a good job. What's your name?"

3

Freelancing and Union Work

▼

Life as a Freelancer

Because of the increasing growth of independent filmmaking, most people working in film production today are freelancers. The term *freelancer* does not mean a person who does not become a full-time salaried employee on a production. It simply means that the individual moves from one production to another. Freelancers may be employed as independent contractors (responsible for reporting their own taxes) or as salaried employees (with the usual deductions in the paycheck). Although there are long-term staff positions at the major studios and television networks, these are the exceptions.

The average duration of a job on a single production depends on what is being produced, the budget, the length of the script, and the length of time for which the individual is being employed. Most commercials, videos, and industrials undergo less than a month of preproduction and less than a week of photography. Features may be prepped for several months, and photography can last from several weeks to nearly one year. This means that most people in the motion picture/television industry regularly experience the nightmare of unemployment.

Two classes of variables determine how many days a year a freelancer works. One variable consists of the duration of the specific position itself, and the other has to do with the person's proficiency and reputation.

Some positions last longer than others. A freelance production manager is usually hired during preproduction and kept on through photography and during some of postproduction. A location manager may be hired only for preproduction, and a grip may be hired for just the actual shoot. A commercial shoot that preps for one week and shoots for two days would give the production coordinator a minimum of 7 days of work, the location scout 5 days, and the grip 2 days of employment on the same production.

But if the grip is well known in the industry and has a lot of experience and contacts, he may do 50 commercials a year. In the final analysis, the duration of a job is helpful in accumulating working days, but the number of job offers is more important.

Although most production people do not work as much as they want, many do approach their goal. However, it takes time to accumulate enough experience and contacts to work continuously. A busy transportation co-ordinator explained it best: "The only reason I am here is because I have gone through the months of unemployment and the moments, days, and weeks of despair. But it is perseverance that counts. I am talented, but what differentiates me from someone who is more talented than I am is that I have the ability to persevere."

Because freelancing plays such an important role in working in the motion picture/television industry we will discuss some of the strategies necessary in order to freelance successfully, including use of the telephone, setting rates of pay, and proper set etiquette.

THE TELEPHONE

The telephone is the lifeline to employment. The sound of a telephone ringing means more to a freelancer in the motion picture/television industry than it does to someone who holds a "regular" job. A ringing telephone holds the promise of work, an old client calling or a new client who has just been referred. Because the telephone provides jobs, every professional in this industry has an answering machine, service, or beeper or some other way of leaving telephone messages or of being contacted immediately. Messages are checked regularly. A lost call is a lost opportunity. This attitude is even more essential in the beginning of a career. When a production manager calls someone she has worked with before and gets an answering machine or service, she will probably wait for a return call. However, when the production manager calls a new person whose expertise is not as highly valued, she will leave a message, hang up, and most probably try the next person on her list. It is important to monitor your calls and to return calls immediately.

RATES OF PAY

When the telephone rings and it is a job, you will first be asked if you are available, and then you will probably be asked a question regarding your day rate (rate of pay). A figure may be suggested by the person offering the job. Getting a call means that someone wants to hire you; therefore the prospective employer wants to agree on a salary. If there is a single phrase that newcomers in this industry should live by, it is these sage words: "thou shall work inexpensively." The best and fastest way to get started in film is to be prepared to accept any job at any rate of pay. If you are new it will be very difficult to convince a department head, production manager, or producer that you are worth $300 a day. However, it is much easier to convince the producer of a student film or of a low-budget independent film that you are worth whatever they have allotted in their budget or mileage money or a reel (a copy of the finished film or video).

Be prepared to work for low or "no-paying" jobs. In this industry, experience is measured by the time spent on the set. The hours of work that you accumulate by paying your dues on low- or no-budget films can be transformed later into good-paying jobs on budgeted films.

One location scout explained that when he started in the industry he asked for $80 a day and got it. Two years later he was making $300 a day. One day a producer offered him $600 a week to do a feature and the scout turned it down. The producer said, "You don't want it badly enough." The location scout explained that of course he did not want it badly enough because he did not need it badly enough.

In the beginning, charge very little because you need every job you can get to accumulate experience. As your experience and reputation grow, so should your salary expectations.

After you have accumulated experience and confidence in your job and developed a list of contacts, you will be faced with the question of how much you should realistically charge. There are industry standard rates of pay for the various jobs (see the career profiles in the following chapters for a partial list and *Brooks Standard Rate Book* [Appendix B] for a complete list). However, most rates are negotiable. A good rule to follow is to charge as much as you can without jeopardizing the job offer. Many producers feel that they get what they pay for. If they are paying top dollar, they feel they should be hiring someone who is at the top of her craft. Be realistic about your own capabilities, and do the best you can at the negotiating table. Producers and production managers have attained their own level of success because they are good negotiators, so be prepared.

SET ETIQUETTE

Another aspect of successful freelancing is knowing how to behave on a set once you are hired. Beyond successfully executing the responsibilities of your position, it is important to exhibit set demeanor. The essence of set demeanor, or set etiquette, is the ability to radiate an outward appearance of professionalism. This is important because it instills confidence in others and tells them that you are well versed in your responsibilities and that you know what is expected of you. It also signals your dedication to the production and your willingness to go the extra mile.

To delineate better what set etiquette encompasses, a producer, gaffer, makeup artist, production manager, and best boy were asked how they would define set demeanor.

Producer

I like to work with people who are conscientious, responsible, and professional, people who will listen to what the director says and not be argumentative or stubborn.

Gaffer

Set etiquette is 10% ability and 90% attitude. It is knowing when to be quiet. It is showing proper respect to the director. Do not sit down, do not stand by the telephone, and do not wonder what is for lunch. Never really relax. Remember that it is a privilege to be working.

Makeup Artist

If you are leaving the set, let someone know where you are so that if you are needed people are not running around looking for you.

Production Manager

Never complain. There are 40 or more people on the set trying to achieve one goal. The etiquette must be all for the shot, whatever it takes. The attitude of "I have been working twelve hours" and whining do nothing but take everyone else down.

Best Boy

Noise factor is critical—you should know the proper time to be making jokes, because everyone likes a good joke, but you must know the time to be quiet. Another aspect is being a hero. For example, if filming is about to begin and something that must be on the set immediately has been over-looked, there you are. You have it, and wham, you can handle it. Go. You have not held up the show at all, and you may have covered for someone else. If you ever hold up the show, that is the ultimate mistake. Mistakes are made that are not visible all the time, but on the set it is critical that you understand your position. Experience equals time on the set. So set demeanor comes from experience. Sometimes you will see people on the set who are very quiet. For all people starting out, this is the safest bet.

Set etiquette is exhibiting the attitude that you are willing to dedicate all your skills and resources to the successful completion of the shot. Good set etiquette is as important to freelancing successfully as having an an-swering machine and establishing the proper rate of pay. According to one key grip, "technicians who have good set etiquette work all the time."

Union and Nonunion

Inevitably when you consider getting started in this industry, you wonder if you need to be a union member. The answer is no. Today, with the large proportion of independent nonunion production, you do not have to be a member of any union or guild to begin working in the motion picture/ television industry. In fact, even if you wanted to join the union for your particular position, you would probably find it difficult in the beginning. As the number of union shows continues to decrease compared with nonunion shows, the unions are having a tough time keeping their current membership employed. And most unions will rarely accept new applicants until all their present members are working.

It is important to understand the distinction between the terms *independent, nonunion,* and *union* production. An independent production is one that is not financed by a major studio (i.e., MGM, Paramount, Warner Bros.) and is produced without a crew that is employed by a major studio. Independent productions may be distributed by a major studio, but the production itself uses a crew composed of independent (nonstudio) free-lancers. However, an independent show may be either a union or a nonunion production. Some "indies" are signatory to the various unions and guilds;

others are signatory to none of the unions but will hire some union members and create a "mixed crew."

The largest union governing the motion picture/television industry is the International Alliance of Theatrical Stage Employees and Moving Picture Machine Operators of the United States and Canada (IATSE and MPMO). Referred to as the IA, this union is affiliated with the AFL-CIO. The IA covers nearly every crew job and also encompasses projectionists, laboratory technicians, and film distributors. Each job category within the IA has its own local offices (or "locals") situated throughout the country.

Another union, authorized by the AFL-CIO to represent the motion picture/television crew members, is called the National Association of Broadcast Employees and Technicians (NABET). The regulations governing the work of NABET members are less rigid than for the IA. NABET has no minimum crewing requirement (the producer can hire any number of crew members), and there are no apprenticeship requirements for membership (unlike the IA). Many people who join a union begin with NABET because it is easier to get in and then later move to the IA. An individual cannot be a member of both unions at once.

A third union concerning crew members is the Teamsters. The Teamsters regulate the motion picture industry drivers, location managers, and other transportation-related positions. Since it is a separate union from the IA and NABET, a producer must sign a separate contract to use Teamster drivers.

Important unions for above-the-line positions include the Writers Guild of America, the Producers Guild of America, and the Directors Guild of America. Each of these guilds sets certain requirements (usually screen credits) for membership and offers symposiums, collective bargaining, and other advantages to their members.

The major studios (i.e., Warner Bros., Disney, Paramount) are signatory to the IA, and therefore all studio pictures are union productions. However, independent producers can decide which, if any, unions they will be signatory to. Many independent producers feel that on low- to medium-sized budget pictures it is less expensive to produce if the crew is nonunion. Today, because more low-budget films are being made than are big-budget films, the trend continues to be in favor of more nonunion (non-IA) independent production.

At some time in your career you may want to join the union. In the past, different locals have, for limited periods, opened their doors and provided apprenticeship and training programs. Today, these kinds of programs are very rare, but it is worth contacting the locals in your area to see if any such programs are planned. Statistically, most nonunion crew members join the union when a union production wants to hire them or when the nonunion show that they are working on is organized by the union. If a nonunion show is organized, the nonunion worker is permitted to use those days towards meeting the union's membership requirements.

Most individuals who are part of the motion picture/television industry have a definite opinion about the unions. These opinions, which usually reflect their status as a union member or nonmember, range from "we owe our high rates of pay to the unions and we need them to maintain those rates" to "the union is busting the back of the film industry and that is why production is running away to all the corners of the globe."

There is no question that historically the unions have been good to the employees of the film industry in establishing benefits for health, pension and welfare, standard minimum pay schedules, and overtime. The question of whether the union is good for you is a personal one that you must examine yourself. However, in the beginning, the question is probably an academic one. The low-budget shows you will be working on will probably not be signatory to the union, and therefore the chance to be a union member will only come after you accumulate experience.

Part II

How Films and Videos Are Produced

The Birth of a Project

As you can see from the preceding chapters, any film or video project is a sizable undertaking. Even the simplest video requires the contributions of a range of diversely talented individuals. Big productions can appear staggering. The "chain of production" may begin months or years before any photography takes place and may end months or years after photography has finished.

The Deal

All major projects begin with a deal; it is the first link in the chain of production. The deal may include any combination of the following elements: the property (script, book, play, song, or creative "concept"), a star or stars, a distribution arrangement, a director, a financier (an individual, a limited partnership, a corporate sponsor, a studio or distribution company), and one person to spearhead the whole project, usually called the producer. The producer is responsible for bringing together enough of these elements and enough of each element to begin production.

Generally on a feature film, the combination of these elements is called a "package." Packaging began during the old studio system when the majors (e.g., Columbia, MGM) already controlled the money, personnel, and many of the creative elements within their studios. Writers, directors, actors, producers, production capital, and distributors were all available to them, and this made it relatively easy to create a package. Today, packaging is still a common practice, but with the advent of many independent productions it has become increasingly challenging. An independent production company must do exactly what the old studios did but without their corporate might or personnel. For the independent producer, frequently the most difficult element to obtain is financing. Often, producers try to attract investors by obtaining the commitment of a well-respected actor, director, or writer to enhance their package's performance value and attractiveness.

Commercials and music videos are usually bid on by production companies. An advertising agency brings the property (the scripted or storyboarded commercial) to the commercial production company, and the agency pays for the production, which, of course, is passed directly to the client. In music videos the product is a musical group's song, and the production company passes the cost of production to the record company and to the artist.

Many educational and instructive programs are financed by corporate sponsors. For these programs, a producer may develop a project and then write funding proposals to prospective sponsors, such as the grant departments of large corporations.

Once the deal is set and the production is financed, the producer can initiate preproduction.

Film or Video?

An important early decision, necessary before preproduction can proceed too far, is the medium on which the production will be shot. Sometimes the medium is dictated from the start by the type of production; other times the producer and other project managers will have the luxury of choice.

Currently there are two methods for photographing and retaining images used by the motion picture/television industry—film and videotape. Film, the older of the two methods, uses a thin sheet of acetate-based material coated with a photosensitive emulsion. A chemical reaction takes place on the emulsion when it is exposed to light. Videotape uses a different method to record. It relies on a wide magnetic tape to retain electromagnetic signals. Visual images are converted into signals for recording and then reconverted into images during playback. Although many people feel that film has a softer, more subtle look, the advantage to video is that it does not have to undergo chemical processing or development. Images on videotape "appear" immediately and require no processing or printing.

Additional differences between film and video include the size of cameras used, light sensitivity, frame size, and methods of sound recording and editing.

The ultimate decision on which medium to use depends to a great extent on the market for the production. Full-length motion pictures designed to be shown in movie theatres are shot on film. A few television shows, commercials, and documentaries are also shot on film because their creators desire the "film look." However, most projects for television—these range from sitcoms and hour-long dramas to made-for-television movies, commercials, and music videos—are shot on video. Some motion pictures and other kinds of productions (how-to, travel, concerts) are shot on video and released only through home video outlets. Many other productions that need greater ease of shooting and speed or simply do not require high production values may also be shot on video. These might include news stories and instructional projects.

Although the early choice of medium is necessary for proper planning, the difference in recording equipment does relatively little to change the fundamental techniques and procedures used in production. Producing a film or video does not involve simply switching on a film or video camera (although most producers probably wish it did). Recording the images and sounds and editing represent only two links in the long chain of production.

Preproduction

Everything that must be readied before the first foot of film or videotape is exposed is done during preproduction. Preproduction may begin hours, days, weeks, or months before actual filming commences. The amount of time allotted to preproduction depends on the project's size, complexity, and budget. Large studio films usually have 4 to 9 months of preproduction. Independent films may have 1 to 3 months. Commercials and half-hour television shows may have two weeks or less, and music videos may have only a few days.

In a start-up situation (not a television series that is producing another episode), the producer hires the production manager (also called the unit production manager or U.P.M.) who is responsible for overseeing the daily logistics of production and for keeping the production on budget and on schedule. If a director is not already part of the deal, then one is selected by the producer. In the case of a studio film (a film being produced by a major studio), studio executives will help choose the director. The property will change and evolve as the director, producer, and writer offer their input.

During preproduction, the producer, director, and production manager hire the department heads and other key people in the crew. The director will probably want to choose a director of photography (D.P.), and the producer may hire a first assistant director (first A.D.). The first A.D. is responsible for running the production on the set and for creating the shooting schedule (usually in the form of production boards), which designate when each scene will be shot and in what order.

Productions are usually not filmed sequentially as they appear in the story. For example, if a movie consists of three scenes ((1) Mike meeting Raphaële at the beach during the day, (2) taking her walking by the lake that afternoon, and (3) driving along the beach later that night), the walk by the lake (the second scene) may be filmed the first day. The two beach scenes may be filmed next, with the daylight scene followed by the night driving scene. This is not done to confuse the actors, who may play their opening scenes last and their last scenes first, but for logistical reasons including the avail-

		EXT. FISHING VILLAGE	EXT. FISHING VILLAGE	INT. STORE	INT. LOUIS' OFFICE	EXT. RAILROAD TRACKS	EXT. BEACH	EXT. BEACH	EXT. BEACH	EXT. SHIP	EXT. NAVAL BASE
Date											
Day or night		D	D	D	N	N	N	D	D/N	D/N	N
Page count		1/8	4/8	3	2	1 4/8	2 4/8	2	6/8	1 2/8	2
Scene numbers		71	100	101 102	83	89 90	16	63 64	50	49	31 32
						91 92					
Title "The Shell"											
Producer											
Director											
Production manager											
Assistant director											
Script dated											
Character / **Performer**	**No.**										
Melville	1				1		1	1	1		
Ernest	2	2	2	2							2
John	3			3							
Diane	4					4				4	
Amie	5	5	5	5							5
Louis	6				6						
Storekeeper	7			7							
Sailor	8		8							8	
	9										
	10										
Extras (#)	11	(20)	(20)	(5)					(10)		(15)
Special effects	12					X					
Special equipment								Jeep			

Figure 1 *Panel from a production board.*

ability of certain actors and locations, budgetary considerations, and overall convenience.

Because most shows are shot out of sequence, the production boards, the master shooting schedule created by the first A.D. for the shoot (or by the production manager in smaller productions), are very important. The production boards consist of multiple panels composed of thin strips of colored paper. The left-hand column lists each cast member, and above this list are horizontal lines that designate the scene number, page count (how many pages of script comprise that particular scene), whether the scene is day or night, the title of the project, and the names of the director, producer, production manager, and first A.D. Each vertical strip on the board represents one scene in the production and the scenes are grouped according to location. The coloring of the strips are used to delineate certain information. For example, blue strips may designate interiors and yellow strips exteriors. Each of the strips is easily removed and reinserted into the panel to accommodate changes and contingencies and to combine various scenes to arrive at the best and most efficient shooting schedule.

On very small shoots, boards may not be prepared at all but the various departments will be informed of the shooting schedule through memos and frequent consultations with the production manager.

On most large productions and in other situations where extra organization is required, a daily shooting schedule is prepared and distributed to the crew. Such schedules generally present information from the production boards in a more convenient form. The complete shooting schedule would be given to the crew at least a few days before filming or taping begins.

When the department heads and other key crew people are hired to do a show, they read the script and do a breakdown. (A breakdown simply means reading the script and looking for certain information.) For a location manager, the breakdown would consist of the number and type of locations mentioned in the script. The key grip is also looking at locations and at the various rigging ideas and camera moves suggested in the script. The art department is reading for a description of the various sets and scenes. The makeup artist is breaking the script down in terms of characters and what happens. Does the mad scientist lose a fight with the game warden? The script says his face is cut and bruised. The transportation coordinator is counting the number and type of picture cars and estimating the number of equipment trucks and trailers that will be needed.

But a script breakdown only offers part of the information that each department head requires. The script says "the family lives in a Victorian house," but is it big or small, with or without columns, new or old? For more specific creative information, the department heads turn to the di-

SHOOTING SCHEDULE

Director _____   Production no. _____

Unit manager _____   Title ___"The Shell"___

Assistant director _____

Day/date	Location	Scene number/set/description	Cast
Tuesday 1st day	Marina Del Rey	Scene 71 ⅞ pg EXT. FISHING VILLAGE Ernest and Amie meet on the dock.	2. Ernest 5. Amie
	Marina Del Rey	Scene 100 ⁴⁄₈ pg. EXT. FISHING VILLAGE Sailor arrives	2. Ernest 5. Amie 8. Sailor
	Marina Del Rey (C.G. base)	Scene 49 1²⁄₈ pg. EXT. SHIP Diane and sailor find the shell.	4. Diane 8. Sailor
	Marina Del Rey (office)	Scene 83 2 pg. INT. LOUIS' OFFICE Melville and Louis talk about the shell.	1. Melville 6. Louis
		End of 1st day 3⅞ total ppgs.	

Figure 2 *Sample daily shooting schedule. These schedules are based on the production boards and can make breakdown and preparation easier for department heads and crew.*

rector. The director will communicate his ideas to the various key crew members, and the crew members will use this information in making their own creative choices. The director may show pictures from fashion magazines to the costume designer and describe how he envisions the "Victorian house" to the location manager. He will discuss with the production designer his overall concept of the production and meet with the other departments that require his creative input.

In addition, the department heads rely on the production boards or schedule to make their decisions regarding rentals and personnel. This information is vital and permits each department to determine accurately what is required each shoot day. For example, if page 20 of the script says

```
                         SCRIPT BREAKDOWN
Title  "The Shell"                    Director _____
Set  Practical _____      (D)N   Day __1__     Pages  1/8
Scene # __7_____
Synopsis __Ernest and Amie meet at the dock._____
```

Cast:	Camera:	Process & Plates:
1. _____		
2. Ernest _____		
3. _____	Wardrobe	Make-up/Hair
4. _____	"Pea jacket"	Wig
5. Amie _____		
6. _____	Electrical/Grip	FX
7. _____	Crane (Chapman)	Smoke
8. _____		
9. _____		
10. _____	Livestock	Sound/Music
11. _____		
12. _____		
13. _____	Communication	Permits
14. _____	Walkie Talkies (20)	ITC (30 secs.)
15. _____		
16. _____	Props/Set Dressing	
17. _____	Fishing pole/Shell	
_____		
_____	Transportation/Vehicles	
_____	∅	

Atmosphere 20   Standing 2   Regular___   Bits___   Welfare worker___   Teacher 1

Figure 3 *A sample worksheet used to break down a script. Different departments would use different sheets or perhaps not use them at all, depending on their specific needs.*

Scene 10   Daylight   Beach

Mike drives to the beach in a new silver Mercedes convertible.

Then page 70 says

Scene 100   Night   Beach

> Mike and Raphaële speed along the road next to the beach in the Mercedes.

Here, the script describes two scenes with one particular car but does not explain if the two scenes are to be filmed during the same 24-hour period. The transportation coordinator does not know if the car will be needed for one day or one month. Only after consulting the boards can this be determined. In this example, the boards tell him that scene 10, the daylight scene, is filmed on March 1 and scene 100, the night scene, is shot on April 1. The transportation coordinator must therefore rent the car for two separate days or possibly for the entire month. (A car rented for a single day and then for an additional day one month later may not look like the same car. If the night scene in the film is supposed to happen on the "same day," any unexpected change in the car's appearance would confuse the audience.)

With creative information supplied by the script and the director and scheduling information supplied by the production boards, and the first A.D. or production manager, each department can prepare for the first day of shooting. The location manager finds the locations with the "right" looks that are available on the shoot days. The art department begins by designing the sets that will be shot first. The technicians put together their equipment packages and schedule their rentals. The transportation department rents the trucks and the picture cars, and the other craftspeople and artisans organize their equipment and tools.

The department heads hire their staffs and provide them with pertinent information about the project. Throughout this process the director, first A.D., and production manager are continually involved. The creative choices of each department are overseen by the director, and the business decisions (i.e., number of crew hired, rate of pay) are overseen by the production manager.

Three final preproduction activities include the preparation of a storyboard, casting, and the technical scout. A producer usually hires a casting director who is responsible for recommending actors for each speaking role and negotiating contracts and salary. Although casting directors recommend actors, it is ultimately the director's decision whom is cast.

Storyboards range from hundreds of multipage panels in feature productions to simple two- or three-page drawings in commercials. Storyboards represent each scene in the script; the camera movement and angle; and they may display props, costumes, and other aspects of the scene. Most storyboards, especially for larger productions, are drawn by a storyboard artist.

One of the final steps undertaken before the shoot begins is the technical scout. At this time the director, together with the D.P., first A.D., production

Figure 4 *Example of a storyboard.* **A,** *Establishing shot. View of a bustling fishing village from the water. There are numerous piers and docks and fishing trawlers roaming the water.* **B,** *View down a pier looking out toward the ocean. On the right, a man fishes. On the left, a man leans against the railing.* **C,** *The man in the previous frame. He is staring out toward the ocean, preoccupied.* **D,** *Medium close-up. A hand slowly reaches into the frame and touches the man's shoulder.* **E,** *Two-shot. The man and the woman face each other. The woman clutches something in her hand.* **F,** *Extreme close-up. The woman opens her hand and reveals the shell.*

designer, gaffer, key grip, and other key crew personnel, visit each location. This is a good opportunity for the director to communicate specific ideas (lighting style, camera moves, production design) with his staff, and staff members can provide suggestions and later review equipment and materials lists to make sure that they are properly prepared.

As the first day of photography approaches, the first location or stage that is scheduled to be filmed may be prepped. The art and construction departments will make the set look exactly the way it is supposed to. The set will be dressed, props will be brought in, and some of the lighting may be prerigged by the technicians. Shooting is about to begin.

Production and Postproduction

That period of the production where actual filming or videotaping takes place has many names, including production, photography, shooting, filming, taping, and principal photography. During the shoot the cameras are brought on to the set and the various scenes are recorded.

There is a subtle distinction between a location and a set. A location (also called "practical location") is any locale that will be filmed outside a studio or stage. The location becomes a set when it is about to experience filming (prepped), is experiencing filming, or has recently experienced filming (strike). A stage may also be called a set when filming or videotaping is taking place. Another way to define a set is wherever the bulk of the crew is located during a shoot.

During photography the size of the crew may range from two people (electronic news gathering, ENG) to over 150 (a big-budget feature film). Of course, this may not be the total number of people attending the production. For example, on a feature film with a crew of 100, there may be 20 actors and 800 extras. This brings the total size of the production, cast plus crew plus extras, to 920 people.

Although a set may appear as a chaotic crowd to an outsider, a strong organization is built in. Maintaining this organization is the responsibility of the first assistant director. The first A.D. keeps the crew informed (i.e., when photography will take place and which scene will be shot next) and

CALL SHEET

Date _____

Location _____ Telephone no. _____ Call time _____ Job no. _____

Clients _____ Telephone no. _____ Product _____

_____ _____ Artist _____

MOS ☐   SYNC ☐   TAPE ☐   16 mm ☐   35 mm ☐   Day _____

Category	Name and home telephone number	Time in	Time out	Category	Name and home telephone number	Time in	Time out
Producer				Gaffer			
Director				Best boy			
Production manager				Electrician			
Assistant director				Electrician			
D.P.				Electrician			
Location manager				Key grip			
Camera operator				Best boy			
1st assistant cameraman				Grip			
2nd assistant cameraman				Grip			
Makeup				Grip driver			
Hair				Swing driver			
Stylist				Set decorator			
Mixer				P.A.			
Art director				Craft service			

Talent name and telephone no.				Agent and telephone no.	Call time		

Call _____ First shot _____ End inv. _____

Lunch _____ Wrapped _____ Exposed footage _____

Camera _____ Mobile home _____ Sound _____

Grip _____ Crane _____ Food _____

Lights _____ Special equip. _____

Figure 5 *Call sheets such as this one further clarify the shooting schedule and provide specific times for when all personnel must report for work. This helps maintain organization during the shoot itself. A call sheet for each day's shooting generally will be distributed the night before. The call sheets are usually prepared by the second assistant director in conjunction with the first assistant director and the production manager. This particular call sheet is fairly basic and might be used in music videos and commercials (note the space for "client information"). A call sheet for feature productions would be similar but would allow for more detail and personnel and perhaps for information on a multi-day shooting schedule.*

makes sure each crew and cast member is at the right place at the right time.

The Shooting Day

The average shooting day lasts a minimum of 12 hours. Depending on the scene, filming can begin at night (in the case of shooting an exterior night scene) or during the day. A daytime shoot usually starts early in the morning. A call time of 7 AM is normal, and certain members of the crew may be expected to arrive even earlier (e.g., drivers to park the equipment trucks, location manager to make sure the location is open, makeup and hairstylist to prepare the actors).

The cast and crew have been given maps to show where the location is and where to park. On many occasions, the parking lot is located a distance from the set. In these cases, shuttle vans are provided to transport cast and crew to the set. The filming location is usually easy to find. Outside the practical location or stage being shot are various vehicles and pieces of equipment including the generator, grip, electric trucks, and motor homes. On arriving, the crew stops by the craftservice table and enjoys a bagel or doughnut and a cup of coffee. Craftservice is a specific position in production and is usually composed of one or two individuals whose main responsibility is to provide a cold breakfast in the morning and snacks and hot and cold beverages throughout the shooting day (or night).

A little after call time, after most of the crew has arrived and has had a bite to eat, the director walks through the location with his director of photography (D.P.) and is usually followed by an entourage comprised of the gaffer, the key grip, the sound person, the camera operator, someone representing the art department, and perhaps the location manager. The director refines the blocking of the shots. ("The actor will come in through the door here and walk up these stairs. I want the camera to look this way and this way and this way.")

The technicians set the shot by bringing in the appropriate equipment (i.e., dolly track, dolly, camera, lights, flags), and if sound is being recorded the soundmixer sets up the recording machine. The art department makes sure the scene is dressed exactly the way it should, and stand-ins may be provided while the crew sets up and lights the scene. The location manager meets with the property owner and explains the shooting schedule for that day, and the first A.D. checks and rechecks that everything needed for shooting that day is either on the set or will soon be there.

When everything is prepared for the first scene, the director is called back to look at the setup. If she approves, then she will say something like "let's try one." The actors are brought out and given the opportunity to do a complete rehearsal (walkthrough or runthrough). The D.P. rechecks light levels while the camera operator and dolly grip practice the camera movement. When everything is ready, the first A.D. shouts "quiet on the set" to quiet the crew and let them know that a take is about to begin. Then the first A.D. says "roll camera." This is a cue for the camera and sound to start recording. When the camera is running, the camera operator says "rolling," and when the sound recorder is running at the right speed the mixer says "speed." A slate is held up by a camera assistant on which will be written the director's and D.P.'s names, the roll, scene, and take number, and the date. The slate is clapped to provide a cue for synchronization of sound and picture, the director calls "action," and acting begins. When the scene is over or when the director decides to abort the take, she says "cut."

Although the director is in charge of all creative decisions regarding the production, filmmaking is a collaborative art, and very often changes are based on suggestions from other crew members. The director may ask the D.P., "What about getting up high on this one?" And the D.P. says, "I think that will look great." The gaffer tells the D.P., "If you are going high I can give you some beautiful light through those top windows." The director tries the shot and it is great, but now the camera is looking at a wall that has not been dressed. "We need a painting or something on this wall," the director says. The art department provides the painting. Suddenly the shot has changed; the angle, lighting, and decor are all new. This kind of flexibility and change is normal on the set. Each shot is a microcosm of the film, and as each shot evolves so does the production itself.

Sometimes there are special difficulties in getting a perfect take from start to finish, and pickups and cutaways will be used. A *pickup shot* can mean either shooting after principal photography is over or reshooting a portion of a scene that has already been photographed. For example, perhaps there is a scene on a basketball court where the actor jumps off the bleachers, runs onto the court, picks up a basketball, and throws it into the basket. When the scene is shot, the actor jumps down off the bleachers and runs onto the court beautifully, but everytime he throws the ball he misses the basket. After several takes, the director may do a pickup by starting the take with the actor already standing on the court and picking up the ball. This pickup will be reshot until the actor makes the basket. It is more expeditious to reshoot the difficult part of a scene than to start from the beginning each time, but unfortunately sometimes the two shots (in this example, jumping off the bleachers and throwing the ball) cannot be edited together without creating a jump in the action. In these cases a *cutaway* will be shot (such as a closeup of the basket or of the actor's face) that will later be inserted in between the two shots. A cutaway allows a scene to continue smoothly by cutting from one shot to another to return to the first shot at a later point.

VIEWING THE TAKES

When using video it is standard procedure to view what is being shot as it is taped or just a few minutes after the scene is completed. It is important to see if the scene worked, if it looked good, if it was framed right and lit properly, and if it contained good performances. However, when shooting 35 mm film it is too expensive to print unwanted takes, and there is not the convenience of on-set review. In film, therefore, the director will select during the shoot the takes he wants to print by saying "print it." These takes are circled on the camera and sound reports and on the script supervisor's log. Later, the printed takes will be viewed during the screening of the dailies. "Dailies," as the name implies, are scenes that have been shot, usually on a single day, that are processed that night and screened the following day. Watching dailies or viewing videotape provides an opportunity for the crew to see how their work appears and allows them to improve their creative and technical approaches to the shoot before the production is over.

Postproduction

Postproduction begins as soon as photography is finished. The amount of time that a film or videotape spends in postproduction is usually longer than the time it takes to do the actual photography and may take even longer than the combined periods of preproduction and photography.

The early stages of postproduction are spent paying final purchase orders and invoices, returning rented equipment, props, and dressing, and allowing the crew to take care of any loose ends. Except for the production accountant and production manager, most of the production staff and crew are now finished with the show.

The remainder of postproduction is spent taking the reels of raw production film or cassettes of tape and assembling from them the finished product. This process is generally termed editing, but it is very involved and may include many different tasks such as adding sound (music, voiceover, looping, sound effects, or sweetening), adding optical and special photographic effects (dissolves, wipes, split screens, miniatures, matte paintings), and adding titles. These postproduction processes depend on the skills and imagination of expert technicians and artisans including the picture editor, the dialogue editor, the sound effects editor, the composer and music editor, and, of course, the director of the production.

Sometimes, more shooting will be conducted during postproduction. Called second unit photography, this may involve pickups, inserts (closeups of specific action), or other shots needed for continuity. Second unit crews are small and may consist of a second unit director, a camera operator and an assistant and perhaps one or two technicians and a P.A.

When a finished product is in hand, postproduction ends.

Part III

Production Jobs and Services

In this section we will examine eight areas or departments in motion picture/television production. Each chapter includes a career profile, a description of the position's duties and responsibilities, interviews with professionals working within the department, and suggestions on how to get started. The techniques and suggestions in these sections will assist you to begin working, and many will be the same methods you will use to stay employed throughout your career. The last chapter in this section examines six support services that production companies rely on and that present the production-oriented entrepreneur with an opportunity to service the film industry. The positions, departments, and services discussed represent some of the fundamental departments and businesses in the motion picture/television industry but do not represent a comprehensive listing of all jobs and services involved in production. These particular occupations were chosen for detailed examination because they provide the best entry-level opportunities.

Keep in mind that you do not have to live in Hollywood or New York City to work in production. With the expansion and decentralization of motion picture production, there are film and video companies in nearly all metropolitan regions: St. Louis, Boston, Albuquerque, San Antonio, Richmond, Washington, DC, Charleston, Miami, Tucson, Kansas City, Dallas, Philadelphia, Cleveland, Nashville, Houston, Minneapolis, Louisville, Tampa, Chicago, and New Orleans.

In addition, there is often a greater opportunity to learn about a variety of film positions outside Southern California, New York City, and the other centers of production. The major production centers tend to depend on a highly specialized workpool—people who are experts in specific roles. Outside these production centers, because the crews and the workpools are smaller, there is a larger dependency on the generalist, the person who can be a camera operator on Monday and the art director on Tuesday.

Finally, it is important to remember that working in "film" does not automatically mean working in feature films or movies. There are production companies that make documentaries, commercials, business films, industrials, educational, instructional, and medical films; there are governmental production offices; there are news and sports show producers; and there are experimental filmmakers. Features generally have the highest budgets and hire the most experienced crews. Therefore, many times it is much easier to begin with the other types of films and videos.

So before you pack up to leave for the Big Orange or Apple, reach for your local telephone book and find out what is being produced next door.

Production Assistant

Job description   Support and assist all other departments in production.

Median income $125 per day.

Work closely with   All members of the crew, especially the second assistant director (A.D.) (on the set) and the production coordinator (in the production office).

Basic requirements   Production kit (paper, pens, etc.), maps, car, organizational ability, high level of motivation.

Employment period   Set P.A.: Principal photography.
Staff P.A.: Preproduction, photography, postproduction.

Duties

The job of the production assistant (called P.A., production associate, runner, or utility) has been called the ultimate entry-level position. The production assistant is a "catch-all" job that supports nearly every other position in production. During preproduction, the P.A. runs errands, delivers copies of the script, types crew sheets, unloads boxes of equipment and furniture, makes maps, telephones crew and cast members to inform them of schedule changes and new call times, and helps out wherever help is needed.

During photography, the P.A. who works on the set takes on a whole new set of responsibilities. These responsibilities may involve working for almost any department. In the morning, the P.A. may help the location manager park crew cars and help craftservice set up their table. Once the cameras begin to roll the P.A. may assist the grip department move equipment. Perhaps the propmaster needs an assistant but does not have one, so the P.A. may work with him or the second A.D. may need a P.A. to escort cast members from their dressing rooms to the set. At the end of the day, when wrap has been called, the P.A. may help the electricians return the lamps to the truck, remove the tape marks from the floor, and clean up the set. The possibilities are endless, and therefore so are the opportunities. A P.A. can learn hands on about many of the production departments in the motion picture/television industry.

Some production companies maintain their own fulltime staff production assistants. These P.A.s are called *runners* if they spend most of their time performing errands. The staff runner position is similar to a mailroom job at a big studio. Both positions offer similar advantages; the runner will be able to learn about the company from the inside. Staff office runners may make runs to the set but will rarely work on the set. The P.A. who works on the set is usually called a set P.A. and is expected to be more knowledgeable about filming than the runner. Many set P.A.s begin as runners.

A low- or medium-sized budget feature, television, commercial, or video production may hire two to six set P.A.s and one to three office runners. Each set P.A. may find himself assisting the same department every day. Maybe the electrical department is short one person, so the best boy electrician grabs a P.A. and has him help load and unload equipment each morning. Or, the set dresser does not have enough assistants, and every day she has the same P.A. help her move furniture.

The most important quality in a P.A., whether on the set or in the production office, is attitude. P.A.s are not expected to know everything about motion picture production, but they are expected to pitch in with energy and enthusiasm where and when they are needed. Resourcefulness is also a valuable quality. When the cameras stop because something is suddenly needed on the set and cannot be found, it is usually a P.A. who is sent on an emergency run. It is important for the P.A. to "pull through" and return to the set with whatever is needed. In the words of one production manager, it is extremely important that the P.A. "makes whatever has to happen, happen."

The job of the P.A. can be both physically and mentally demanding. Although the rate of pay is considered low by industry standards, the P.A. is compensated in the form of education.

Although most P.A.s are not in a union, some may join the Director's Guild of America (DGA) if they want to become A.D.s. The DGA allows production assistants to join if they are working on a video production that is signatory with the DGA. In addition, the National Alliance of Broadcast Engineers and Technicians (NABET) has a production assistant category but recommends that P.A.s first get employed by a company signatory to NABET before paying their membership fees to join the union.

Because the job of P.A. is known as an entry-level position, a good P.A. who is liked by the crew is in an enviable position. This P.A. can ask questions of the seasoned pros, learn about whatever interests him in production, and make valuable contacts in those areas. It is not a coincidence that the two P.A.s interviewed for this chapter have already moved up the ranks and were no longer P.A.s when we spoke with them.

Interviews with Production Assistants

John Grant began his career in the motion picture/television industry as a P.A. Today he is a production manager who works on major national commercials. His credits include Pepsi, McDonald's, Budweiser, Bartles and James Wine Cooler, Burger King, and Nissan.

As a production manager who works a lot on location, you must be approached by people who want to break in as a P.A. What do you tell them?

A lot of people ask me outside of Los Angeles. Most of the time I turn it around back to them because they have no idea what movie industry work is about. I look for, first of all, somebody who knows the terrain and the area like the back of his hand because that is very important to me. If somebody says I need to get an old-fashioned wheelbarrow and this guy can say, "My father is an antique restorer, and he may have one," that is important. Now when they come to Los Angeles I always encourage them to hit the streets. Many times I would see a movie being filmed, and would stop and look around and see if I knew somebody on the crew. Then I would be introduced to somebody else, I would get to know a few faces, shake a few hands, say hello to a few old friends, and then I would move along on my merry way, and it is all part of the decorum of this industry. It is networking. Even with only six months of experience you may end up working with the same people. It is very important to establish what we call accounts. That means you set yourself up with somebody who really likes you and will use you every time they work.

What kind of special talents or abilities should a P.A. possess?

The attitude must be there: "Sure, anything you need. Yes sir, right away. I will get that. I will take care of that." That attitude will take the P.A. to so many more levels. It is being polite and talking to people of superior level with respect. Say, for instance, you are told as a P.A. to go over to the wardrobe stylist's house and do this, that, and the other thing. And you go over to the stylist's house and tell her that you do not know why you are there and you start complaining. Well, that stylist will call up the producer and say, "Why did you send me this bimbo?" It is all attitude. Say instead, "OK, I am working with the stylist. Sure, I will go over there. What is her name? Where does she live? Anything else? Do you want me to call you from there?"

It is all for the project. It is important to be project oriented. If you have a lot of outside commitments, sometimes they have to be put aside for the time being. Sometimes you are not going to get home until 9, 10, or 11 PM. That is a given. If you are not willing to accept that, do not take the job. If you do not have the tools, do not take the call. If you do not have a car, then I am not going to hire you. In New York it might work but not on the west coast. And it should be a dependable car. The last thing a coordinator wants to hear is, "I have got brake problems and I may need a transmission; I am not sure if I can make it." I do not want to hear that. That is part of what you bring to the job—past experience, other jobs, a decent car that runs.

I look for a P.A. who is eager, willing to learn, and not afraid to take on new responsibilities and to help everyone out. I will turn over paperwork to a P.A. who says, "Can I do the production reports? Can I help you with

DAILY PRODUCTION REPORT

No. of days on episode including today					
Travel	Holidays	Idle	Rehearsals	Work	Total

No. days scheduled	No. days actual	Ahead/behind

Title _____   Production no. _____   Date _____

Producer _____   Director _____   Assistant director _____

Date started _____   Scheduled finish _____   Estimated finish _____

Sets _____
_____
_____

Crew call _____   First shot _____   Lunch _____   Estimated wrap _____

Dinner _____   First shot after dinner _____   Estimated wrap _____

Company dismissed: at studio _____   on location _____

Film use	Gross	Print	NG	Waste	Rolls
Previous					
Today					
To date					

Film inventory: Start _____   Additional _____   Total _____

Script scenes/pages	Scene/pages	Minutes	Setups	Added
Previous				
Today				
Total				

Cast			Work time		Meals		
Cast	Character	Makeup	Call	Dismiss	Out	In	Signature

Comments: (Delays, explanations, etc.) _____
_____
_____
_____
_____

Figure 6 *Daily production reports help key personnel monitor the project's status. Such reports are generally assembled by the second assistant director and reviewed and approved by the production manager.*

the contracts?" That will help the P.A. to grow even faster and further. I look for somebody who is eager and willing to jump right in and pick up where she sees that she is needed.

So with a production report, a P.A. does not have to know exactly what that is but if she's willing to learn, she'll be taught what a production report is.

Exactly. Sometimes it is a sink or swim situation—here is your chance. You have 800 extras out there—get them together and move them into the set. That is a sink or swim situation for a P.A. He can do it or he cannot. It is not that any one situation makes or breaks a P.A.'s career chances, but then again, if there is an opportunity to take on more responsibility and learn, that experience will help all the way through the rest of your career, because you will gain hands-on experience from seasoned veterans.

What other qualities are important?

It is important to be able to meet and work with new people very quickly and to understand that if you approach people with the right attitude, you can accomplish anything. Commitment is very important. A major minus for people is if they quit jobs or walk away from something unfinished or if they leave a bad taste in somebody's mouth. Commitment is a must in a production person.

Are the hours long for a P.A.?

Yes. Often the P.A. is the first one to arrive on the set and the last one to go home, but that pays. That commitment and extra effort show that you are willing to take that extra step, and it will take you that much further later down the line.

In addition to having a car and the kind of attitude you describe, what other tools does a P.A. need?

A P.A. should carry a production kit with him composed of three or four pads of paper in different sizes, several pens and pencils, marking pens, and a production directory and map for the area. If you are in the middle of nowhere and someone says, "Draw me a map from here to there" and you do not have a pad of paper or a ruler or a pencil or a marking pen. . . . It is all part of what makes your polish and your shine a little more bright. If you know that you are going out to a location in the predawn hours, take a flashlight. It is part of the job. Think about it, it is all common sense.

The job of P.A. has been called the "ultimate entry-level job." Is it true that it can lead to many of the other jobs in production?

Absolutely. It is all that one break. That one foot in the door, and then the sky is the limit. You can go anywhere you want to go. You can do anything you want to do. And getting the first foot in the door is where you have to be faster and smarter than the other person. My drama teacher at Hollywood High taught us that the door to success is labeled "push." Remember that.

Willy Mann began his career in the motion picture/television industry as a P.A. Currently he works as a best boy in the grip department. He has worked on commercials and feature films.

Is the hierarchy in production more rigid than in the corporate world?

More people have a chance of getting to the top and impressing their ideas on those who are at the top in the film business than in corporate business. I think a lot of people end up in the film business because it satisfies their desire not to be part of the corporate structure.

How did you get started as a P.A.?

I moved out here specifically to get into this business. I came out with a pocketful of money that I had made working in Alaska on the railroad during the pipeline boom where I had gone after college. I did not even know what a P.A. was. After a while, I got involved with some first-time projects where the people could not be too discriminating about whom they hired in terms of experience.

What special abilities does a P.A. need to do his or her job well?

Humility, and understanding that you are there ultimately to assist the director in making the movie. As a P.A., your ability to understand that and to support the production in every facet is extremely important. It is also important to understand that the job of P.A. means that you are ultimately there to assist whoever needs help, and that requires a person with a multifaceted personality. You must be flexible.

What do you mean by flexible?

First, you must realize that you have been very fortunate to get the job, and the time you spend on the set you can observe other people and see what areas you want to go into. You are in a position ultimately to help everyone. As an art department person you won't get near putting a hand on the camera, but as a P.A. there may come a time when you are helping the camera operator, doing a slate or something. That is the beauty of it, the broad experience that can be gained. And if you can capitalize on that, you can take it anywhere you want to go in this business.

Getting Started

> *"Your time is so directed as a P.A. that the fact that you have never been on a show can mean very little."*
> —*A former production assistant.*

THE RESUME

To get a job as a P.A. you should prepare a good, clean, simple resume. Explain who you are and briefly include any background material or previous work experience that exemplifies your ability to work hard under pressure. If you have any film- or video-related experience, include that of course. Do not state that you want to be a director, producer, writer, or any other lofty production goal. No one wants to hire a P.A. who is slumming it. Of course, you can have those goals and, the truth is, your experience as a P.A. will help you achieve any future production-related objective. But for the sake of sounding dedicated with the single-minded determination associated with accomplishing a single goal, keep the directing off your resume for now. Reiterate your willingness to work hard and any other special abilities that would make you a valuable P.A. Perhaps you know the town like the back of your hand, and no one can get from A to B as quickly as you can. Perhaps you can type 60 words per minute.

Put yourself in the place of the person reading your resume. As a producer, production coordinator, or production manager who is hiring a P.A., you want someone who is motivated, willing to work hard, and can take direction well. Any part of your background that illustrates your competence in these areas will be beneficial. Along with the resume, send a cover letter that states, "I will follow this up with a telephone call in a few days after you have had a chance to review my resume." That says that you will take two steps instead of one to get the job. Showing your motivation now gives your prospective employer an indication of the kind of motivation you would use later on a production.

THE TRADES

The film industry trade magazines and newspapers are your information source for those companies that are starting up or are currently in production. A production company may have several staff runners, but once filming commences the company may have to hire three or four set P.A.s. When a trade announcement indicates that a production is starting up, this is the time to contact the office. Call up the production company and try to speak with the production coordinator or manager. If you are told all the positions are filled, find out how long their shooting schedule is. There is a "Murphy's law" that says that even the best-planned productions can always use an extra hand. If the company has just begun filming and plans to shoot for several weeks or months, mail in or drop off your resume and contact the company every couple of weeks. There may soon be a "new opening."

COLD CALLS

"Call everybody" was one former P.A.'s advice when she explained how to get started. Pick up your local production directory and call all those companies that may hire a P.A. or staff runner. This includes not just production companies (commercial production, music video, industrial and educational) but also still photography studios, advertising agencies, cable production companies, special effects companies, record companies, and stages. A step in any of these directions is a step toward working in production. When you call, ask for the coordinator or manager and stress the same qualities written in your resume. (See the section on resumes.)

VISIT SETS

This technique must be conducted with the subtlety of an Indian snake charmer, but if properly done it can be very effective. First it must be stated that when a production company is shooting at a publicly accessible location (e.g., a beach, a city street, a park), the company is not there to entertain questions from the public. With that said, it is acceptable when cameras are not rolling to ask a member of the crew to direct you to the production manager. If you are asked why, tell her that it is to find out if they need an extra hand. If you are told that the P.M. cannot see you now, thank the person very much and ask the name of the P.M. If you have that, later when you call the production company, you will know who to ask for. However, do not be too surprised if your first question is answered with a hand pointing to a someone sitting on a chair with a stack of paper in her arms, looking very worried. Approach cautiously, excuse yourself for interrupting, smile, and introduce yourself. Be courteous and to the point. "Hello, my name is so and so . . . I was wondering if you needed another runner or P.A. I am willing to work hard and inexpensively, and I am very familiar with the area." Whether or not you get the job now, it is important to make a good impression. If the answer is no or I will know in a few days, ask if you can leave your name and phone number or send her your resume. Shake hands and leave. You have just made a valuable contact; stay in touch with her. (To find out where film and videos are being shot in your area, contact your local motion picture liaison office. For a list of these offices, see Appendix B.)

GET LISTED IN PRODUCTION DIRECTORIES

Have your name and telephone number listed in the P.A. category of your local production directories. The telephone number you list should be connected to an answering service or machine, and messages should be checked regularly.

A NOTE ON THE INTERVIEW

When you are interviewed by the production coordinator or production manager, do not expound about Hitchcock's influence on film noir. Instead, reiterate those aspects in your resume that got you the interview. You're motivated, hardworking, a good listener, and willing to take direction and to go the extra mile. When you get the job, do all the things you promised you would, and you will be on your way.

Hair/Makeup Department

Job description   The makeup artist is responsible for beautifying, enhancing, and altering the appearance of
actors through the use of makeup. The hairdresser is responsible for beautifying, enhancing, and altering the appearance of actors
through the maintenance and creation of
hairstyles.

Median income   Makeup artist: $1500/week
Hairdresser: $1500/week
Assistant (makeup or hair): $900/week

Work closely with   Makeup artists and hairdressers work closely
together and with the costumer, production
designer, and director.

Basic requirements   Makeup artist: Experience, makeup kit.
Hairdresser: Experience, license to cut hair.

Employment period   Preproduction and principal photography.

Duties

Makeup and hairdressing are both concerned with an actor's appearance. The makeup artist is responsible for enhancing and altering the appearance of the actor's face and body and balancing skin tones for lighting purposes. The hairdresser (or hairstylist) has similar responsibilities, including beautifying or enhancing the appearance of the actor's hair and altering its appearance to suit the look of a specific character role.

For example, in a battle scene, a soldier is shot and receives a surface wound to the scalp. His head is bleeding, and his hair has been torn. There is blood all over his chest and on his coat. He pulls himself through the mud on his elbows and tumbles into a trench. Then, the soldier takes a grenade from deep inside one of his pockets and throws it at the enemy in a last-ditch effort to save his life.

In this scene, the makeup artist would be responsible for the dirt and blood on all exposed areas of the actor's body. If the production has a body makeup artist (most union productions do), then the body makeup artist and makeup artist would work as a team. The makeup artist would apply the fake blood and dirt to the head, face, and forearms, and the body makeup artist would be responsible for applying makeup to the other parts (e.g., the chest). The hairdresser would provide the actor with his disheveled and torn hair and might create a wig for this effect. The costumer would supply the soldier's clothes (uniform, coat, boots, etc.) and would be responsible for any blood or dirt on the costume. While the makeup artist (and the body makeup artist) are concerned with what is literally on the body, the costumer is responsible for any makeup applied to the costume or wardrobe. Finally, the grenade would be provided by the property master. As discussed in Chapter 14, the property master is responsible for any prop that is movable or specifically mentioned in the script.

Makeup artists and hairdressers determine the kinds of looks they will create for the actors based on reading the script and discussing ideas with the director and production designer. In a production where the actor is playing "himself" or "herself," the treatment involves enhancing the actor's own look. However, when a character is being created, such as a 19th century aristocrat, a Roman gladiator, an alien, or anything that is not mainstream or contemporary, the makeup artist and hairdresser must rely on the director's vision, the brief descriptions offered in the script, and their own research and interpretations to apply the appropriate makeup and styling.

The greatest challenge facing the makeup artist and hairdresser is that in most productions, scenes are shot out of sequence. There may be a western brawl scene, for example, and an actor walks into a saloon, gets into a fight, gets a black eye and a cut lip, leaves the saloon, steals a horse, and rides away. The interiors may be shot on a stage in a studio, and the exteriors may be shot in a real western town. During the beginning of the shooting schedule, the production goes on the road and shoots the exterior of the saloon. The makeup artist and hairdresser give the actor his "normal look" for the scene when he walks toward the saloon and steps inside. Then he has the fight, which will be filmed later, and the makeup artist and hairdresser give the actor his "after-the-fight-look," consisting of black eyes, cut lip, disheveled hair, etc. With this look, the actor is filmed leaving the saloon, stealing the horse, and riding away. One month later, the company moves to a stage to shoot the interior of the saloon brawl. The actor's "normal look" must be recreated as we see him entering the saloon (this time from the inside). As the scene unfolds and the fight takes place, the actor must be given his "after-the fight-look." The interior stage sequence will later be edited together with the exteriors, and therefore the actor's makeup and hair, both before and after the fight, must perfectly match the looks he had in the scenes shot weeks before. The makeup artist and hairdresser use Polaroid pictures and notes to recreate both looks identically. A mistake could cost the production a great deal of time and money. There is a story of a western film where, in mid-production, an actor's scar was placed on the wrong side of his face. All the sequences filmed with the misplaced scar had to be reshot at a cost of hundreds of thousands of dollars.

In addition to recreating identical looks after weeks or months have passed, another challenge facing the makeup artist and hairdresser is maintaining the same look throughout the day of shooting. If an actor appears in three scenes and in all three scenes she has the same look and the scenes are filmed on the same day, then the makeup artist and hairdresser must keep the look identical and fresh throughout the long (twelve hours or more) shooting day. This requires constant maintenance and touchup. Of course, if the look changes back and forth between scenes, then the makeup artist and hairdresser are faced with the challenges of continuity described above.

The number of makeup artists and hairdressers used in a production depends on the type, size, and budget of the shoot. A film with big "stars" would require more than one makeup artist and hairdresser. Also, productions with period looks or "out of this world looks" require the expertise of many individuals. In *The Wizard of Oz* nearly 60 makeup artists were hired to create the look of the munchkins.

BEGINNING OF SCHEDULE
Exterior of saloon is filmed

"Normal look"
for walking toward saloon

"After the fight look"
for leaving saloon

ONE MONTH LATER
Interior of saloon is filmed

"Normal look"
for entering saloon

"After the fight look"
for leaving saloon

Figure 7 *Makeup continuity. When films and videos are shot out of sequence, makeup continuity becomes increasingly challenging and important. In this illustration the looks before and after the fight must be consistent with corresponding makeup applied a month before.*

When there is more than one makeup artist or hairdresser on a show, the first one hired for each department is the department head (also called the key) and is responsible for overseeing all the others within their departments. The key makeup artist and key hairdresser have the ultimate responsibility for making sure that each character looks the way the director imagined. On smaller productions, the hairdresser may be hired under the makeup artist and is considered the assistant. On very small productions, the makeup artist and hairdresser may be the same person.

The makeup artist and hairdresser have very early call times to prepare the actors with their "looks." Depending on the complexity of the makeup and styling required, the makeup artist and hairdresser may arrive from one to several hours before the rest of the crew. A 7 AM crew call may mean a

4, 5, or 6 AM call for makeup and hair. Because actors generally see the makeup artist and hairdresser first, before anyone else sees them, there is a strong psychological element to the job as well. Sometimes actors are tired and feel that they do not look right. It is the job of the makeup and hair people to put the actors at ease and assure them that they look great.

In the past, makeup artists were self-taught, but today many graduate from professional makeup schools. Most hairdressers are licensed hair professionals who have spent years practicing their craft in salons.

Interviews with Makeup Artists and Hairdressers

> *Howard Smit has been a makeup artist for nearly 50 years. His credits include* The Wizard of Oz *and* The Birds. *Today, Mr. Smit is the business manager for IA Local 706, the Makeup Artists and Hairstylists union in Los Angeles.*

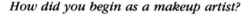

How did you begin as a makeup artist?

I had a slight background in makeup. I fooled around with it when I was a kid in Chicago, but it was only a hobby. As a matter of fact, when I got hooked in makeup, just prior to *The Wizard of Oz,* I was in my last year of law school. So I quit my last year of law school and diligently put all my efforts toward makeup and learning. I will never forget the early days when I had just started—I was so frustrated because I was getting nowhere work-wise and I said to one of the old timers, "How in the hell do you really get going and become a makeup artist?" And he looked at me and said, "You just get in there and keep doing what you are doing and make up your mind that you are going to do it and nobody is going to stop you." And he was right. In those days we were all self-taught and -trained.

Is it the same today?

No. A lot of makeup schools exist today.

I understand that makeup artists and hairdressers often have very early calls.

We must do our work prior to shooting. The hours are horrendous. For a regular 8 AM call, usually the makeup artist and the hairdresser have to be on the set by about 6 AM. And if we have anything working like toupees or beards or mustaches, not only are we the last ones to leave but we can be there for an hour or an hour and a half after everyone has wrapped. We are there cleaning those hairpieces and getting them reading for the following morning.

How do makeup artists get proficient at what they do?

You learn the basics in a book, then you start doing it on people, and you find that if you have a flair for it—it works. But the person who has the ability, the touch for being a makeup artist also has to have the stamina. He or she must go out there and work and work and work until it is the way it should be. We have all tried to add a little and lend a little to build the makeup profession to where they do some very beautiful and subtle work today.

What advice would you have for someone starting as a makeup artist?

It is not easy. It is not handed to people on a silver platter, but if they want to really work at it, it is all there. You must first and foremost learn proficiency in your profession; then, if you have the talent—go.

Jane Galli has been the makeup artist on many features and television movies of the week, including ABC's Street of Dreams, *NBC's* Combat High, *and HBO's* Vietnam War Stories.

How did you begin working?

I worked for free for a long time. You can always find people who will hire you for free, and you are able to practice. In the beginning you should never turn down a job no matter how small it is. Even if it's $50 for a 14-hour day, never turn down a job. You will learn how to take direction and work as a team, and that is extremely important.

Did you go to makeup school?

Yes.

Was it beneficial?

Yes. Makeup school teaches you what products are available and how to apply them to create different effects. You get to experiment and try again until you get it right. And getting it right in school and practicing until you feel comfortable is what school is for, not on a job where you have very little time. Makeup artists must have knowledge in a variety of areas, including beauty and corrective makeup, laying a beard, making a bruised eye, and creating first, second, and third degree burns. You get to practice and do things that you would probably never do again.

What kinds of tools and materials does a makeup artist need?

You need a great deal. You have to be ready for anything. If the director wants tears, you have to be ready with a menthol blower, which is a little machine that blows menthol mist into the actor's eyes and makes them tear. In your makeup kit you carry everything from fake blood to different sizes of fake eyelashes. I did a show once where there was a fake doll, and the director suddenly decided he wanted blonde hair on the doll. I carry about ten different colors of hair in my case along with an assortment of mustaches and beards and eyebrows, so I was prepared. You just never know when you will need something.

When you begin work on a new show, what is the first thing you do?

First I break down the script, which means reading the script and writing down notes on the characters and what happens to them, and then I will meet with the director. I will also negotiate for prep time. It is not unusual to have a week or less to get all your stuff together for the shoot, get it all loaded in the trailer, and get the trailer set up, so when you walk in there the first morning of shooting you are ready to go.

Michelle Buhler has worked as a hairdresser and makeup artist on a number of feature films. Her credits include Communion, Tap, *and* The Serpent and the Rainbow.

▼

First I want to ask you about the job of the makeup artist. Who decides how a character will be made up?

It is a joint effort. In a film both the actor and the director will have their vision of what the character looks like. Sometimes the character looks very much like the actor and sometimes not. So the three of us get together, the director, the actor, and myself, and we discuss how this character would wear his makeup, how he would wear his hair, would he be very glamorous, or would he be very down to earth. We also consider the character's life-style, what the movie is saying, and the look of the film.

Is the job of hairdresser very similar to that of makeup artist?

People are funny about their hair. Actors have much more set ideas about their hair than they do about their makeup, probably because they have been doing their hair since they were teenagers. So sometimes it is harder to break through that barrier of saying to an actor, "You might like this hairstyle, but the character would more likely wear this other hairstyle." It is harder to get that through to them than with makeup. Sometimes you may have to talk an actor into accepting the hairstyle for her character, and with all the cutting and coloring, it is also a more permanent thing.

Where do the hairdressers who work in the motion picture/television in-dustry come from?

The people who have worked and who have a lot of experience in salons make the best hairdressers. The main difference between working in a salon and a film is that in a film you are working closely with a lot of strong egos and personalities. In a salon, it is usually the everyday person, and although you may have to know how to work with their personalities, it is on a much softer level. So on a film you must act very confidently so that the actors feel secure in your hands. If you are doing something special, like a period piece, you must do a lot of investigation and research so that you are both confident and knowledgeable.

So there are things you would do on a film that you might never have done in a salon?

Sure, you have to study laying lace wigs and things like that, and that takes practice and a lot of work. That is something you generally would not get to do in a salon. But anyone who has had the basic training in a salon of haircutting is in good shape because they have cut about 10,000 heads. So that is the best basis to come from, and then it is just a matter of learning the more psychological aspects of dealing with actors and, of course, learn-ing to do proper research and preparation.

The psychological aspect is very important?

Being knowledgeable and confident is extremely important. I have seen very good makeup and hair people get eaten alive because they may have backed down or given an impression that they did not have the knowledge. People can be insecure in this business. Therefore, if they do not see you coming from security that ruins it for you, even though you may be great at what you do. So you have to know how to relate to people, work on their level, and be confident. You have to be able to "read" people rather quickly. Even when you go on an interview with a producer, you have to be able to read what she is looking for, and that just comes from confidence and experience.

How would you recommend someone begin as a makeup artist or hairstylist?

The best way to get started is to apprentice with someone and assist people. You can take seminars to learn all the special things you need to know, but there is nothing like being on a set to learn how to act and see what must be done.

Getting Started

Nearly every makeup artist and hairdresser who forged a successful career in the motion picture/television industry had previous experience in the field. Hairdressers should be licensed, and makeup artists should have attended makeup seminars or, even better, a makeup school. On large productions, you can expect to get started by assisting the makeup artist and hairdresser. On smaller shows, you may have the opportunity to begin working in the key position itself.

READ THE TRADES

Films and videos currently under production are listed in the back pages of the industry trade newspapers and magazines. Contact the low-budget productions and student films that are "crewing up" and impress them with your capability to do the job and your willingness to work inexpensively. Generally, the more money a company is willing to pay, the more experience they expect from you. In the beginning, when you do not have much experience, keep your salary expectations low.

Often, student films offer a copy of the finished film or video as compensation for your participation in the production. These "reels" are valuable and can be used to show potential paying employers examples of your work. In addition, film students do graduate and go on to larger productions, and they will remember you.

Always take pictures of your work in low-budget and student films. These photographs are important components of your resume and portfolio and proof of your ability.

WORK AS AN ASSISTANT

Contact makeup artists and hairdressers. Send out letters of introduction, resumes, and business cards. Let the working professionals know who you are and that you are ready, willing, and able to work and would appreciate any opportunity to be an assistant. Stress that your salary expectations are low. Listings of makeup artists and hairdressers can be found through local film industry directories. (See Appendix A.)

WORK IN A SALON

Hairdressers can get a job in a good salon that caters to clients in the motion picture/television industry. Meeting industry members in this fashion presents a great opportunity to show them the kind of work you can do.

PREPARE A PORTFOLIO

One of the most important tools that hair and makeup people use to get hired is their "book" or portfolio. The portfolio consists of photographs of actors or models who are wearing hair or makeup styles prepared by the artist or stylist. A book can be compiled in a variety of ways, including doing someone's hair or makeup for free in exchange for a chance to photograph the subject.

ATTEND SEMINARS

Motion picture/television seminars for makeup and hair are offered occasionally and announced through trade publications and makeup schools. By attending these seminars you have an opportunity to learn new techniques and to meet working professionals in your field.

Transportation

Career Profile

Job description	The transportation department is responsible for everything that rolls in a production, including all equipment vehicles used behind the scenes and all picture vehicles used in front of the cameras.
Median income	Transportation captain: $1250/week Driver: $750/week
Work closely with	The transportation coordinator (head of department) and the captain work with all department heads. Drivers interact with most cast and crew members.
Basic requirements	License to drive multiaxle vehicles and passenger vans is a plus.
Employment period	Coordinator and captain: Preproduction and photography. Driver: Photography.

Duties

The transportation department is responsible for getting to the set and maintaining everything that rolls and has a motor. This includes picture vehicles, the cars, vans, and motorcycles used in front of the camera as part of the scene and all equipment vehicles, the motorhomes, trailers, prop, grip, and electric trucks used by the production behind the scenes.

The head of the transportation department is called the transportation coordinator. The coordinator is responsible for renting all the vehicles necessary for the production, hiring the drivers (in a nonunion show), and overseeing every aspect of transportation related to the production.

The transportation coordinator is brought in during preproduction to determine the type and number of equipment, picture, and specialty vehicles that the show requires. The coordinator does this by meeting with the production manager and the various department heads, reading the script, and studying the shooting schedule or production boards.

The production boards are extremely important because the script provides only limited information on the picture vehicles needed (e.g., "the detective pulls up to the house in a blue two-door sedan with Florida plates"). The boards state how many days the car will be used and if those days are consecutive.

With input from the art department, the transportation coordinator will rent or otherwise arrange for the various picture vehicles to be provided for the shoot. Depending on the script, such picture vehicles can include anything from an antique Model T to a Lotus Esprit sports car.

As the beginning of principal photography approaches, the coordinator working on a nonunion show (a production that does not use Teamster drivers) will hire a captain and a cocaptain, if one is needed. On a union show, these individuals are supplied to the production from the union's roster. Although the coordinator holds the ultimate authority in the department, the transportation captain and cocaptain are in charge of transportation on the set and for moving and parking the equipment, maintaining the vehicles, and shuttling actors and crew members. This leaves the coordinator free to spend time in the production office to locate and rent vehicles and respond to changes in scheduling.

In a large show, there may be more than 20 picture vehicles and over 400 linear feet of equipment, including four or five motorhomes or trailers, numerous 5-ton and 10-ton trucks, and an assortment of specialty vehicles such as cranes, scissor lifts, and backhoes. In a production that uses a large

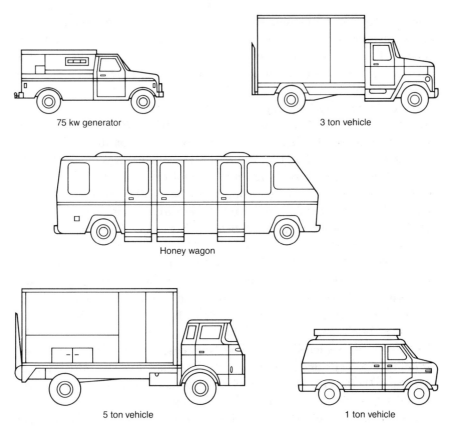

75 kw generator

3 ton vehicle

Honey wagon

5 ton vehicle

1 ton vehicle

Figure 8 *Transportation equipment vehicles. Examples of some equipment vehicles driven by members of the transportation department during a production.*

number of picture cars, a picture car captain may be hired to prep, gas, maintain, and oversee the delivery of the picture cars to the set.

The drivers are brought on as soon as they are needed. In the case of a "road show," in which the first day of shooting is scheduled to take place at some remote location, the drivers may be hired days or weeks before the filming begins to drive the equipment to the location. The total number of drivers that is hired depends on the production's size and budget and the total number of vehicles that will be used. Nonunion shows tend to hire a smaller number of drivers than do Teamster shows.

Once hired, each driver will be designated an equipment vehicle or picture car that he is responsible for getting to the set and maintaining. If a single location is being filmed for an extended period, some of the drivers may be let go (on a nonunion show) once they deliver the equipment, and a skeleton crew will stay to maintain the vehicles and shuttle the production personnel. However, most shows today use a variety of locations, and it is not unusual for all the production's vehicles to be moved to a different location every day.

Drivers work very long days and are expected to arrive at a location a minimum of one hour before the first crew call. If the first person scheduled to arrive on the set is the makeup artist at 6:15 AM, then the driver of the makeup trailer is expected to be there at 5:15 AM. This gives the driver time to park, balance, and prepare the vehicle for the day's use. If the crew wraps at 7 PM, then all the equipment might not be carried and stowed in the truck until an hour or two later. If the drivers are moving their vehicles to the next day's location the night before, they conceivably could not be finished with work until 10 or 11 PM.

Throughout the day, drivers are also expected to drive the shuttle vehicles, usually passenger vans, between the set and wherever the crew's and cast's personal cars are parked. An additional responsibility includes making "runs" when people or things must be delivered to the set or to the production office.

Driving some vehicles requires special licenses. Anyone with a standard automobile driver's license (called class 3 in California) can drive a two-axle truck. However, to drive a truck that has more than two axles requires a class 1 license, and a shuttle van driver should have a class 2 license, which is similar to a chauffeur's license. (Check with your local Department of Motor Vehicles.)

The transportation department is also responsible for any specialty equipment that may be required on the set. Water trucks needed for wetdowns are used often. Camera cranes and scissor lifts used to lift a camera and one or more operators high above the ground are also frequently requested. Because of the varied needs that the transportation department services, good drivers quickly learn how to drive and operate everything from a 40-foot motorhome to a backhoe to a Ferrari.

Smaller productions that go on location may not have a full-fledged transportation department. These smaller-sized productions such as commercials and music videos may hire drivers and rely on the production manager or coordinator to do the job of the transportation coordinator.

Interviews with Transportation Coordinators

> *Mike Shepherd has been a union transportation coordinator for nearly 28 years. His credits include* Winds of War, The Tracker *and various IMAX productions.*

I understand that in the driver's union, the Teamsters, there is a ranking system for all drivers, transportation captains, and cocaptains.

Drivers are listed on a roster that is based on seniority. The drivers with the most experience, the most seniority, get hired first.

How did the roster system evolve?

The seniority system began in this business years ago when RKO and Columbia and the other majors did large western films. At that time there were about 500 drivers in the whole industry. When one of the majors fired up five features before any other studio did, they would hire the cream of the crop, the very best guys. This left the other studios with the dregs of the business. So the studios got together and said, "Listen fellows, this has got to stop, we are going to have to share the wealth." So they started the seniority system.

How would you describe the job of the transportation department?

You are in charge of everything that has a motor and moves in front or behind of the camera. You are in charge of all movement, not only of the personnel but also of the equipment.

As the transportation coordinator on a show, how do you decide how many and what types of vehicles will be needed?

You have to take a look at your cast and crew list. You have to know, for example, the contracts that the actors work under. Some require a 13- to 35-foot motorhome with all the amenities. Some of them have the "Jim Backus clause," which means "me too." If the other guy gets a circus tent for a dressing room, then I get a circus tent. If the other guy gets a 4-foot by 8-foot room, then I get a 4-foot by 8-foot room. Those are some of the things you have to know to put your equipment package together. You have to talk with the key grip, the set decorator, and the cinematographer. If, as you read the script, you realize it is a heavy prop show, then you know that you have to sit down with the property master and say, "Listen, instead of a 5 ton we are going to put you in a 10 ton because of the large number of props that you are going to have." So it is done basically through a one-on-one situation between the transportation coordinator and the department heads. After it is settled as to what they feel they are going to need, then you sit down with the production manager and find out whether it is in the budget.

What kind of special abilities, beyond the ability to drive, do you look for in your drivers?

They have to be able to address people. We move everybody on the production from the person who picks up the trash to the top stars in the business. So you always have to have the proper attitude.

> *Rick Rollison has worked as a nonunion transportation coordinator on many television shows and independent feature films. His credits include* Sledge Hammer *and* Two Moon Junction.

▼

I know my way around trucks, I have done some driving, and I am mechanically inclined; would that be somebody who you think would do well in the transportation department?

I like to hire good drivers who are multitalented because in nonunion transportation you have to do a lot of things. I am always looking for someone who can work on different kinds of cars and who knows vehicles, equipment, generators, cranes, scissor lifts, backhoes, forklifts, and bulldozers because there are calls for all those things and more. But there are other skills I am looking for, including someone who has a personable nature. I do not care if a guy is overqualified and he can do anything—if he is an insufferable jerk I cannot use him. I would rather train someone who is amiable and politically inclined and teach him what he does not know. Also, the hours are long and tempers can be short. I need someone who can not just go the extra mile but also can be nice about it and maintain a professional attitude.

How long are the days?

Regular in-town truckers work eight to twelve hours a day and then go home and lead a regular life. In the transportation department of production, you are there one to two hours before anyone comes to the set and you are there one to two hours after they have all gone home. Fourteen- to sixteen-hour days are the norm. Eighteen- to twenty-four hour days are not unusual.

It sounds as if dedication is a pretty important aspect of the job.

Absolutely. I need people who, when they make a commitment to work for me, will carry through. I want people that I can rely on and trust. People who, if faced with a problem, will know how to attack it instead of having to halt in the middle of their task. A lot of people do not have the ability to walk through a problem and solve it. In the field of transportation you will constantly be challenged by problems. The production will throw curves at you right and left, and if you are not good at solving problems then you are not good at doing your job.

A lot of the time stress goes along with the job. Stress is a big factor. You have to find your own way to deal with that. For me, it is humor. I can always teach a driver to drive, but I cannot teach him to tell a good joke.

What would you tell a new person, from out of town, who wants to work in transportation?

Good luck! It is dicey to hire people from out of town because they do not know the area. You cannot send them on a run without telling them how to get there. That wastes time. If I am doing a big show, I may bring someone from out of town on and will watch him, and if I think he will work out, I will invest the extra time to train him. Plus, he must be willing to pay his dues, because no one ever really stops paying dues. I do a lot of hard shows, and in this industry you have to prove yourself on every show, or you will not be hired back. Hard work begets hard work.

What sort of qualifications do you look for in a captain?

They have to be good drivers. That is the first ability that they should have. The other qualities a captain should have include being a good tactician, a good dispatcher, a diplomat and a professional in everything he does. He is in a position of power, and he deals with all the other departments in the production.

Where do the picture cars used in a film come from?

Every piece of equipment that I am required to get I am expected to get at the best possible deal. Now if they want a car, for example, a Mercedes 560 SL, first we will try to get it on a promotional basis. A company may provide it for free for coverage in a film. If we cannot get it that way, then we have to rent it.

How did you get started in this industry?

The natural progression was I began as a P.A. on commercials, working like crazy, doing a thankless job, lugging coolers, and setting up craft service. Eventually, I started working with an art director. Then I went into props, and I prop-mastered two shows. Then I art directed a feature and a cheap television series, and I found out very quickly that it was not what I wanted to do. Then I had the opportunity to drive for a friend. He got me for $250 a week, and I paid my dues and worked like a monster for him. That is what you must do. You have to prove yourself to somebody. Once you prove yourself, you will get more and more responsibility. I worked cheap for about a year or so. Finally, I was selected as one of the transportation captains on *Sledge Hammer.* Later the transportation coordinator left to coordinate another show and moved me in to coordinate the remaining 11 episodes. And I have been working steadily ever since.

Do you still enjoy transportation?

I love it. It is still challenging for me, and the money is getting better and better. I still find it fun. And once it stops being fun I will channel my energy into some other area.

Getting Started

The International Brotherhood of Teamsters represents the drivers in the motion picture/television industry. In recent years the numbers of independent nonunion productions has increased while the number of union productions has, at best, remained the same. Most Teamster locals are busy keeping their own members working and have very little room on their rosters for new people. Therefore, working on independent, non-Teamster shows provides the most accessible way to get your first job as a driver in the film industry. Later, as you gain more experience, you may have the opportunity to join the union.

CONTACT FILM OFFICES

Today, there are government-sponsored film liaison offices located in most major cities in the United States. These offices provide location information and are a source of support services to production companies coming into the area to film. Production companies often contact film liaison offices and request the names of equipment suppliers, production facilities, and crew members, especially location scouts and drivers. Contact the offices in your area and ask if you can get your name and telephone number on a transportation list or crew roster that the office provides to production companies. (A list of film liaison offices appears in Appendix B.)

OBTAIN LISTING IN PRODUCTION DIRECTORIES

There are production directories published for every major film and video production center in the United States and the world (see Appendix A). Contact these publications and get your name and telephone number listed in the transportation sections. Some of the directories print only the names of experienced drivers and may require you to submit your resume. In this case, it is important that you have held some driving-related job in production. Use one of the other suggestions discussed in this section to get your first job to get listed.

READ THE TRADES

The various entertainment trade newspapers and magazines list productions that are beginning to shoot or are "crewing up" (see Appendix C). Contact these productions either through the mail or, if possible, on the telephone. Refer correspondence to the transportation coordinator. If you find that one has not been hired yet, try to find out when the coordinator will start and make it a point to contact the office at that time. If you are told that there is not a transportation coordinator and there is no plan to hire one (perhaps it is a small production), ask for the production manager or the production coordinator. The responsibility to hire drivers falls on their shoulders without a transportation department head. Once you are in contact with the person who has the authority to hire you, you may be told that the production has all the drivers it needs. Leave this person your name and telephone number anyway and even go down to the production office and meet her, if that is possible. Often production schedules change once filming commences, and a production that had plenty of drivers suddenly finds itself short-handed. If your telephone number is in the right hands, you may get a call days or weeks after the shoot began.

CONTACT NONUNION TRANSPORTATION COORDINATORS

Union (Teamster) transportation coordinators must obey the union guidelines, which stipulate that they only employ union drivers. However, independent nonunion productions and nonunion transportation coordinators can hire anyone they want. Nonunion coordinators can be found in most production directories and through referrals at some of the truck and trailer rental outfits. A letter of introduction that mentions any transportation-related experience you might have (e.g., special licenses) and a business card with your telephone number should be mailed to these coordinators and followed up with a telephone call. If they have nothing for you now, ask them if you can call back in a couple of weeks. In this business, jobs can begin very quickly. In addition, if one transportation coordinator does not have any work available for you, he may recommend another coordinator: "Try so and so. I'll give you his number. I hear he is just starting up a show." Follow all leads.

CONTACT PRODUCTION COMPANIES

Contact the production companies in your area. This includes commercial and video producers and industrial, educational, and of course television and feature production companies that shoot on location and rely on drivers to get their vehicles to and from the set. Of course, these companies already have "their" drivers, but you would be surprised how often they are not available because they are on another job. Your letter or business card sitting on their desk may get you in.

VISIT OPEN SETS

If you are in an area where a lot of filming takes place on the streets, next time you see a shoot make it a point to stop and talk with the drivers. Most drivers are extremely friendly and personable. Do not be afraid to ask them about work or whom you should approach for work. Maybe they will point to a guy in a blue parka and say, "He is the captain, go talk to him." That is a great opportunity. When you speak with the transportation captain, do not forget to leave him your name and telephone number.

10

Grip and Electric Departments

▼

Job description   The grip department is responsible for moving the camera and creating shadows. The electrical department is responsible for setting up and powering the lights and other equipment requiring electricity.

Median income   Beginning grip or electrician: $150/day
Experienced grip or electrician: $250/day

Work closely with   Grips and electricians work closely with each other and with the camera department.

Basic requirements   Ability to problem solve, use tools, and work with a team of people.

Employment period   Photography, although department heads usually start during preproduction.

Duties

The electrical and grip departments are the backbone of the motion picture crew. The electricians are responsible for powering everything electrical except the recording of sound. They provide power to the lights and run the cables. If it has a plug on it, it is usually the electrician's responsibility. The grip department supports the electricians and the camera department. Grips are responsible for securing the lights, moving the camera, creating the shadows, and constructing any special rigs such as attaching a camera to the sideboard of a bus.

Although the two departments work closely as a team, the following examples will highlight their individual duties.

An electrician can plug in a light and illuminate an entire room, but without diffusing filters or flags, the light will make the room as bright as the sun. The grips will provide the appropriate filters and flags (outside and in front of the lamp) to diffuse the light and to create shadows. In addition, if the camera moves on a dolly or crane or similar device, the grip is responsible for laying the dolly track, attaching the camera, and physically moving the dolly or crane.

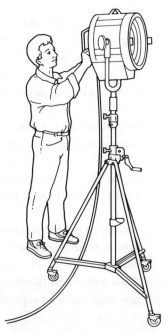

Figure 9 *Electrician positioning a light. Electricians are responsible for powering and positioning the lamps (lights).*

Lighting directions usually begin with the director of photography (D.P.) and are conveyed to the gaffer (the head of the electrical department). The gaffer relies on his crew of electricians and the grip department to provide the proper illumination, diffusion, and shadows that the D.P. requested. Camera moves are also conveyed from the director to the D.P. to the key grip, but sometimes the director will communicate her ideas directly to the key grip or to the dolly grip (the grip who is responsible for moving the dolly or crane).

The hierarchy in the electrical and grip department breaks down in the following way:

	Electrical	*Grip*
Department head	Gaffer	Key grip
Department administrator	Best boy electric	Best boy grip
Special position		Dolly grip
Workers	Electricians	Grips

Both the key grip and gaffer are hired before filming commences and, during preproduction, have similar responsibilities for their respective departments. Both scout locations with the director and D.P. to formulate lists of the kinds and amount of equipment that their departments will need to create the shots that the director and D.P. plan.

During filming, the gaffer and key grip will often position themselves near the D.P. to quickly provide the kinds of lighting and framing that the D.P. requests. The D.P. might say, "Give me some shadows criss-crossing over the guy who will be standing in the corner of the room right there." The gaffer will have her electricians power and position the lights, and the key grip will tell his grips to position some shading devices in front of the lamps to create the criss-cross shadow effect.

Later, the director might tell the D.P., "I want to follow this actor as he leaves the corner of the room, goes to the table over there, and sits down." The D.P. would explain the shot to the key grip, who would in turn tell his grips to bring in dolly track and then show them where to lay it.

Grips are also responsible for moving the dolly or raising or lowering the crane during a "move" (when the camera is connected to the dolly or crane for a shot). These "moves" require subtle timing, strength, and precision. Often marks must be hit at specific moments, positions must be changed fluidly, and the dolly must move at a precise speed.

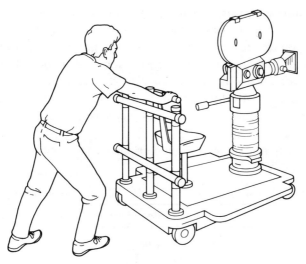

Figure 10 *Dolly grip. Dolly grips, within the grip department, have the special responsibility of moving the dolly. Often the pace and framing of a scene is a result of the dolly grip's pacing, rhythm, and precision.*

Sometimes, where the responsibilities of the electricians end and the grips begin can become very subtle on large productions and union shows. In these cases the electricians are responsible for attaching any filter or gel directly to the lamp, while the grips are responsible for any filters that are in front of but separate from the light.

Both the grip and electrical departments have a position called best boy, and their responsibilities are similar. Both best boys (best boy electrician or second electrician and best boy grip) are the administrators and supervisors of their respective departments. The best boy electrician is responsible for all the electrical equipment and all the people who work under the gaffer. The best boy grip supervises the grip department and is responsible for their time cards. Both best boys are responsible for seeing that every piece of equipment that left their respective grip and electrical trucks during the shoot are returned at wrap. Although only the second in command, the best boys' administrative responsibilities leave the gaffer and key grip free to work closely with the director of photography. Supervision takes a back seat to creative involvement.

In addition to moving the camera and creating the shadows on the set, the grip department has additional responsibilities during any kind of specialty shot. Sometimes a light or camera is attached to the side of a helicopter or a bus or inside a speeding plane. The grips are responsible for rigging or attaching the camera. In those instances, the grip department and, specifically, the key grip is responsible for safety. Therefore, rigging, whether it be a simple mount on a dolly or a complex one, where a camera is attached to a stunt plane, must always be fixed, safe, and secure. As one key grip explained, "You must be able to walk away from any rig and know that it will not fail, it is not going to fall. If it does, you may kill somebody."

Often mounts and rigs are created on the spot. Suddenly the director may decide to attach the camera to the sidecar of a motorcycle for a point-of-view shot. At those moments, it is the grips who pull out the plywood and the two by fours, drill the holes, bolt the screws—and "voila!"

Interviews with Grips and Electricians

> *Robert Studenny has worked as a key grip and dolly grip on national commercials, music videos, and feature films. His credits include U2, Michael Jackson, Pepsi, McDonalds, and the feature film* Two Moon Junction.

As the key grip, you work closely with the director and the director of photography?

Yes. The director says, "This is what I want. The guy is going to walk out of here. Then he is going to do this and then stop here." As the key grip, I will go and do my work. I will place the camera and then show it to the director of photography. Then the D.P. will work with the gaffer and light the scene. Then we will have a rehearsal and have the actors play out the scene so the D.P. and director can check the lighting and I can see how to move the camera. Filmmaking is one of those crafts that really is a group effort. If D.P.s and directors respect their crew and the crew respects them, they will get a good film every time. A crew that likes the D.P. and director will do almost anything for them.

How are most jobs awarded—through networking or if the right resume lands on the right desk at the right time?

With networking the company already knows who you are when you walk in the door. They have already called around and asked who you are. If they do not know you, then take in a resume. Or you can go in and tell them what you have done. You might not even need a piece of paper. If you mention films that people know about, then they know that you have done the job before. And then it comes down to money and if they can afford you.

What advice would you have for people who are starting out?

Work on every show you can. Accept every job offer even if the pay is low. Hustle and keep humble and understand that you are the low one on the totem pole. Watch people, and when you see someone who knows what he is doing, ask questions. In this industry you have to get in to learn your stuff. You cannot learn it unless you are working. It is the only way to learn it right, and most technicians will teach someone who is hustling for him. They will show him everything they know.

> *Willy Mann has worked as a best boy and grip in features and commercials. His credits include the CBS miniseries* Dream West *and the movies* Johnny Be Good *and* Dead Heat.

How would you describe the job of a grip?

Grip work is not creative in the same way that cinematography is creative. However, it is equally diverse because there are countless ways to handle the various challenges that you face. You do not spring from the head of Zeus as a qualified key grip and know how to handle everything. You grow into that with time and by learning from others. If you have never been on a set before, it can be overwhelming. Not only can you not figure out what you are supposed to do, you are just standing there staring at the actresses. You almost have to have some sort of production background.

What job should you start out at?

All the jobs in film are very interesting, and if you are not there to see them in action then you really do not even know what goes on. So take any job that gets you closer to the set.

Let us say you get a job as a grip. What would you be doing the first day on the set?

In the beginning a lot of it is schlepping, carrying equipment to and from the set. You watch the other guys. With time, you will begin to understand how all the tinker toys fit together. It is a learning process.

How would you recommend to begin as a grip?

One way is to hook up with a guy who does a lot of rigging. He can bring you in when filming is not actually taking place, and you can learn that way.

How did you begin?

I began with a trial by fire on a feature after I had worked only two days as a grip for free on an American Film Institute student film.

Once you begin working, how do you continue to work?

It is word of mouth and something that snowballs with time. The more different jobs you have, the more people you work with, the more exposure you get. Without a doubt, your first job will lead somewhere. You are going to meet someone on that show whom you can call next week and say, "By the way, if you need any help give me a call." But it is important not to get discouraged. You cannot take it as a personal failure if you are not working. Just about everybody goes through times of unemployment in this industry.

> *Alan McKay has worked as an electrician and a gaffer in the motion picture/television industry for many years on both nationally televised commercials and feature films.*

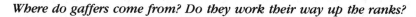

Where do gaffers come from? Do they work their way up the ranks?

Many of the older guys are sons of gaffers and electricians. Others have worked their way up from menial grip and electrical jobs. I went to Navy Film School, trained as a cinematographer, and thought I wanted to be a D.P. Later, I found that the electrical department was a real easy place for someone who is serious about work to find work. Nobody likes electricity. Nobody else really wants to do it. So electrical positions prevail.

As a gaffer, you are the head of the electrical department on the show. Can you give an example of what you do?

The cinematographer says, "Look, this is the idea; I would like a shadow to cut across the guy's face to look like there's a lampshade up above." So then I say, "Great, I know how to do that." I will get the light going, then I will turn to the key grip and say, "Will you put a cutter on this thing? Would you put a topper on it so that the light stops across the guy's eyes so it looks as if he is sitting beneath a poker shade?"

A gaffer is also usually brought in during preproduction?

Yes.

What are you responsible for before principal photography begins?

During preproduction I order gels, lamps, and the other electrical equipment. I determine how many feet we will need of what kind of cable and the number and types of lamps. A lot of that can be determined by scouting the locations. I will also prepare a list of consumables, stuff that will be used up such as gels and wire.

As a gaffer, do you hire your crew or does the unit production manager (U.P.M.)?

The gaffer picks, the U.P.M. hires, and the production company makes the call.

How would you describe the job of the best boy?

The best boy, or second electrician, is a supervisor. Generally, electrical crews are composed of four or five people. He's the boss of those people. As a gaffer, I cannot be bothered with how many extension cords they need. The best boy is actually physically responsible for all the electrical gear, all the people who work under the gaffer, their time cards, and their morale. This way the gaffer can concentrate on working with the cinematographer. The best boy is also responsible for power distribution. He must go around constantly and make sure everything is balanced out and that we are not drawing too much power.

What special abilities does someone need to work in the electrical department?

Strength is important because there is lots of physical work, lifting heavy lamps, etc. It is important that you are a team player because no one does it all, and if you want to work your way up the ranks it is helpful if you have a creative eye. A good gaffer will say to the electricians, "See that light coming through the window over there? Give me more of that." He won't say, "Go get 'X' light and put 'Y' diffusion on it and gel." So I would rather have a guy who can think for himself. I like to be able to say to him, "See that light coming through the kitchen window, make it come through the dining room window too." So it is important to learn to acquire ability and technique by watching.

Getting Started

RENTAL HOUSES

Grips and electricians who are just starting out spend most of their time transporting equipment back and forth between the set and the equipment truck. Therefore, the most important quality for a new grip or electrician is knowing the different names and functions of the equipment. "Get me a c-stand, a sand bag, an apple box, and a flag," can sound confusing if you do not know what it means. Therefore, there is a tendency to hire "new guys" who know something about the tools and equipment.

One of the best ways to learn about the equipment and meet prospective employers is through working in a rental house. Grip and electrical rental houses are the largest sources of equipment. Grips, gaffers, and electricians are constantly in and out of these houses examining new pieces of equipment, filling out rental orders, and off-loading after a show.

As an employee of a rental house you will learn the names and functions of most of the equipment and will have the opportunity to impress working grips and electricians with your knowledge. Then it becomes a matter of letting them know that you are available when they need an inexpensive grip or electrician to fill out their crew. Rental houses are listed in the various production directories. (See Appendix A for a list of production directories.)

ATTEND WORKSHOPS AND SEMINARS

Motion picture/television workshops and seminars specializing in the technical crafts are held throughout the year. Many of the people who attend these programs, both students and teachers, are currently working in film or video production. Therefore, there is an opportunity to learn about your craft from some of the best in the business and to make important contacts.

Film-related workshops and seminars are often listed in the display and classified advertisements in the various trade magazines. (See Appendix C for a list of film and video industry trade publications.) Also, colleges and universities with film departments occasionally sponsor their own.

FREEBIES

The best way to get a job as a grip or electrician is to have some experience in one or both of these areas, and the fastest way to get experience is to work for free. Low- or no-budget independent productions and student films needing crews are listed in the back of the various trade magazines. Often, these productions have little or no money to pay and try to "crew-up" by hiring as many freebies as possible.

Although there might not be any immediate financial reward, there are other valuable forms of compensation. You receive on-the-job training. You have the opportunity to meet other filmmakers, many of whom will go on to bigger-budget shows and who will remember you. Some productions offer their freebie crew a reel at the end of production, so now you have a finished piece of work that you can show paying employers and say, "I worked on this." Sometimes there is a recoupable cash payment that comes in the form of, "When the film sells or the pilot is accepted, we will pay you."

Again, the bottom line is that it is much easier to get hired if you already have experience, and experience equals time spent on the set. By working freebies, you accumulate that time.

RIGGING AND WRAPPING CREWS

Often when a large set is being constructed (rigged) before filming or torn down (wrapped) after filming, a large number of workers are assembled to expedite the process. Because no filming is taking place, the experienced technicians who supervise the rigging or wrapping crew will have more of an inclination and opportunity to teach an individual who wants to learn.

There are many ways to get rigging and wrapping work. Here are two suggestions:

1. Contact set construction companies and offer your services inexpensively.

2. Contact experienced grips and electricians. They often get calls to do this kind of work but may turn some offers down. However, if they have

a telephone number of someone who wants to do it, they may pass it on to the person who needs a crew.

CONTACT GRIPS AND ELECTRICIANS

Ultimately the biggest source of potential employment is other technicians. Grips and electricians work in teams or crews usually composed of from three to seven individuals in each department. An excellent way to get started is to send these individuals a letter introducing yourself and to follow up with a telephone call. Remember to treat these professionals with proper respect and decorum, and they may put you to work. (A listing of technicians is available in most production directories. See Appendix A.)

Locations

Career Profile

Job description   Responsible for finding, photographing, secur-
                        ing, and managing practical locations for film-
                        ing. Practical locations may include roads,
                        houses, and any other locations filmed outside
                        a back lot or studio.

Median income   Experienced location manager: $250 + /day.
                        Location scout: $125 to $250/day.

Work closely with   Production designer, art director, production
                        manager, director, director of photography.

Basic requirements Resourceful, familiar with community, person-
                        able. Car, camera.

Employment period   Location scout: Preproduction.
                        Location manager: Preproduction,
                        photography.

Duties

The location department is responsible for finding, securing, and managing all locations filmed outside the studio or back lot. There are two positions in this department—location manager and location scout. The location scout's primary responsibilities include finding (scouting) the locations, photographing them, negotiating the rental fee with the property owner, and obtaining the proper government film permit. The location manager does all the above and, in addition, manages the location during filming by getting the production company into and out of the location without incident.

Location managers are usually very experienced scouts who specialize in doing larger productions that require the presence of a location expert during filming. In addition to finding and securing the locations, larger productions need someone to "pave the way" for a film company by alerting neighbors, arranging for crew, cast, and equipment parking, and following up after the filming to make sure everything has been left the way it was found. On smaller productions this is also important, but, unfortunately, many production companies view the location manager's job as a luxury they cannot afford and therefore hire a location scout only to find the location. If, however, the production company maintains the scout's services as a public relations person during the actual filming, then the location scout becomes the location manager. Therefore, the fundamental difference between a location manager and location scout depends on the activity. If an individual is hired only during preproduction to find and secure the location, then that person is considered a scout. If the individual is hired for both preproduction and photography, then the person is a location manager.

To avoid confusion (a location manager is always a location scout but a location scout might not be a location manager), the term *scout* will refer to any discussion of preproduction responsibilities, while *location manager* will refer to responsibilities during filming.

The job of the location scout has become more important in recent years. In the past, many productions were filmed in studios or on back lots. Today, the growth of independent production and the higher costs of filming on the few back lots that remain have created an increased dependence on "practical," or real, locations. Practical locations include everything outside a studio or stage, from airplane hangers to subway stations, from mansions to liquor stores.

Immediately after being hired, the location scout is given some information about the locations needed for the production. The scout may be given a storyboard (in the case of a commercial), a "concept" (in the case of a music video), or a script (in the case of a television show or feature-length production). After getting a preliminary idea of the locations from the written materials, the scout meets with the production designer or director to get additional input on how the locations are envisioned.

Next, the scout is responsible for finding and photographing several choices, or "options," for each location needed. For example, a scout may be called in to find a large ranch house for use in a commercial. The director explains that the house should have a picket fence, big rooms, high wood beam ceilings, and a pool in the backyard. The scout pinpoints the best neighborhoods to find this kind of house and scouts those communities by driving up and down the streets. When the scout finds a house that has the "right" exterior, she will knock on the door, explain the project to the owner, and ask permission to see the rest of the house. If the house looks "right" (i.e., large rooms with wood beam ceilings, backyard with a pool), then the scout will take pictures. The scout will also discuss the availability of the house on the shoot date and the approximate rental price and leave with the owner's name, address, and telephone number. Now the scout has one option. A good scout will try to find several so that the production designer or director can choose from a variety.

The photographing of each "option" is done with a 35-mm camera and color print film (although there are occasions where Polaroid or video cameras may be used). The choice of lens is usually up to the location scout, although many ask the director if she has a preference. (Most recommend 28 mm or 50 mm.)

Photographs of the locations are taken from a variety of angles, usually from front, side, and reverse, and accurate notes on the address and contact for each location are kept. The pictures are processed by a one-hour photography laboratory and are boarded. Boarding means pasting the pictures of each location into an individual file folder and marking it with the address of the location, the telephone number, and whom to contact.

Sometimes a director may choose the location from the first group of options presented; other times, he may hone his ideas further and send the scout out several times, each time with more specific instructions.

A location scout can be asked to find almost any practical location in any environment: a trestle bridge over a river, a building with a view of suburbia, or a location that can be faked as something else, such as a California beach that looks like the Bahamas, a cafe that looks like it is in Paris, or a hotel that looks like the 1920s. Exactly how a location scout finds the locations depends on the scout's experience and resourcefulness. All potential sources of information are used, including government-sponsored film location offices, the yellow pages, friends, other location scouts, and businesses that may have a connection to the location sought. For a location that looks like England, the scout may call the British Chamber of Commerce. For a hi-tech restaurant, the scout may contact the local restaurant association. The scout brings to every job the knowledge that if she does not know where a particular location is, somebody else does.

Once a location is selected, the scout usually takes the director, the production designer, the D.P., and other key crew members to the location so they can see it for themselves. If the director selects the location, then the scout alerts the property owner (individual or company) in the case of private property, negotiates the rental price, has the property owner sign a location release, and finds and secures parking for the crew and the equipment. In the case of public property (e.g., park or beach), the scout contacts the appropriate government office and is informed of the appropriate use fee. In addition, the scout may be responsible for securing the necessary government film permit, required for both public and private property.

If the scout is kept on during the shoot, then as the location manager, she will act in a public relations capacity, standing between the community and the production company. The location manager is responsible for making sure that the production company gets the shots it needs and that the filming experience is a positive one, with minimal inconvenience to the private property owner and the neighborhood.

The managing aspect can often be very challenging during large productions or when something special is filmed such as a stunt or action sequence. Sometimes with 100 crew members working in a house on a suburban street and 300 or more feet of equipment out in front, it appears as if a regiment has invaded. In anticipation of this, the location manager has alerted the neighbors and explained the project to minimize the "what the heck is going on" response. The manager is also out there during the shoot answering questions from the production company ("Is there a phone we can use? Where is the bathroom?") and helping the neighbors ("I think one of your trucks is blocking my driveway and I cannot get out"). Both the production company and the community rely on the location manager. If a problem arises, such as a police officer's telling the company that it cannot do what it has planned, then the company says, "Get the location manager." If there is a problem such as a complaining neighbor, then it is the location manager's responsibility to listen and try to satisfy such complaints. A good location manager knows what the production company will do and has received all the necessary permissions before the day of filming.

LOCATION RELEASE

To _____   Date _____

_____

_____

For valuable consideration received, you hereby grant to the undersigned permission
to go upon your property located at _____

together with access to and from said premises, for the purpose of

and/or erecting and maintaining temporary motion picture sets and structures, and of photographing
said premises, sets and structures and/or recording sound for such scenes as lessee may desire,
and/or for such other purposes as lessee may desire.

The undersigned warrants that he is the owner or the agent of said premises; that he is fully
authorized to enter into this agreement and has the right to grant lessee the use of said premises
and each and all of the rights herein granted.

Lessee may take possession of said premises on or about _____ and
may continue in possession thereof until the completion of all photographing and recording for
which lessee may desire the use of said premises for the proposed scenes, including all retakes,
added scenes, changes, process shots, etc.

Lessee agrees to pay as rental for said premises _____.

Lessee shall leave said premises in as good condition as when received by it, reasonable wear,
tear, damage and use of said premises for the purposes herein permitted, excepted; and lessee
shall have the right to remove all of its sets, structures and other material and equipment from said
premises.

Lessee shall own all rights of every kind in and to all photographs and recordings made by it on
or about said premises and shall have the right to use such photographs and/or recordings in any
manner it may desire without limitation or restriction of any kind.

Lessee hereby agrees to indemnify you and hold you harmless from any claims and demands
of any person or persons arising out of or based upon personal injuries and/or death suffered by
such person or persons, resulting directly from any act of negligence on our part while we are
engaged in the photographing of said photoplay upon your premises.

Approved and accepted   Starting date on or about _____

by _____   _____
                                                                  Producing company

                                                   by _____
                                                                  Authorized agent

Figure 11   *Sample location release. The location scout or manager is usually
responsible for negotiating permission to use a site.*

Interviews with a Location Scout and Manager

> *George Herthel is a location manager with many years of experience managing commercials and feature films. His credits include* D.O.A., The Karen Carpenter Story, *and* Promised A Miracle.

▼

How would you describe the job of the location manager?

It is like being the point man for a political candidate. You are right out there in front. You deal with the public and government enterprises as well. It is basically choosing (filming) sites, negotiating contracts, obtaining permits, and working with the police and government officials and making sure that the neighbors and whoever else is affected by the company is happy. It is important to keep everyone happy. It is a fine line between what the company wants to do and what the neighbors and the cities allow it to do. We always run on the outside of the envelope. We get very close to breaking the rules, but we never break the rules.

What kinds of expertise does a location manager need?

In preproduction you are basically satisfying the director's and the production designer's needs as well as adding your own creativity to it. Sometimes to get the right style of a house you have to go into a difficult area where it is not easy to get a movie company in and out. That is your responsibility, to get the company in and out of those areas with the minimum fuss. If you cannot park the trucks where it is adequate, then you cannot shoot there no matter how good the house looks.

How can you learn or acquire these skills?

The ability to negotiate can be an acquired skill, but some people already have that. Once you begin to shoot, the public relations aspect is the hardest part.

How did you get started?

I went to the University of Arizona in Tuscon and studied film. Then I went to Europe for three years and traveled around and raced automobiles. Then I came back here and started doing stunt driving for some of my old sponsors. I thought I might as well put my degree to work because I could not race anymore. I ended my career when I crashed in the German Grand Prix, so then I had to go to work in the real world.

When you scout, how do you actually find the locations you need?

In the beginning I called everybody I knew. . . . I use the film commission. . . . I do a lot of research on my own. I will pop myself in the car for a couple of hours on a weekend and drive around, and whenever I see things that I think will work for something, I make notes. My whole garage is full of all the pictures that I have taken since I started doing this.

> *Mike Paris is a freelance location scout who has worked on commercials, music videos, and independent feature films.*

Do you consider scouting hard work?

The job itself is pretty straightforward. You have to find the locations that the script calls for. The only difficulty comes with the amount of time that you are given to find them. You are working for the director but you are being employed by the producer. The director wants the best, the most beautiful, the most interesting looking locations. If the director could have it her way, she would probably like to have fifty location scouts out looking for the same thing. But the producer is paying the bills. Not only is he not going to hire fifty location scouts, he is only going to give the one scout he has hired a couple of days to find the location. So on one side you get a director wanting to see more, and on the other you get a producer wishing you would find the location as soon as possible. This is more true with commercials and videos than with features.

Once you find a location, how do you convince the homeowner to let you into the house to take pictures?

You present yourself honestly, explain the project, and maybe even give the person an idea of how much the company has budgeted for the location.

"Hi, I am so and so. I am working on so and so. I am looking for a house, and I think this may work. The shoot day is so and so. And the company has approximately this much money to spend." At that point most people are interested, unless their house gets filmed all the time or they have heard horror stories about film companies pillaging whole neighborhoods.

What if someone says no?

Then it is up to your sales ability, how convincing you can be. Most good scouts and managers can talk their way into nearly any property.

Getting Started

WORK AS AN ASSISTANT

Contact experienced location scouts and managers and offer your services as an assistant. As an assistant you will have the opportunity to learn from an experienced professional how to scout, take pictures, negotiate rentals, and obtain government permits. To get a job as an assistant:

1. Use production directories to call production companies (independent feature companies, television, commercials, videos) and ask to speak with the location scout or manager. If they do not have a scout/manager on staff, ask for the names of the people whom they usually employ. (Appendix A lists the various production directories.)

2. Contact working location scouts/managers by obtaining a list of location managers from your local film liaison office. Most film permit–liaison offices print a list of location scouts and managers working in their area. (See Appendix B for a list of film liaison offices.)

Often on low-budget independent features, a location manager is hired without an assistant. On these shows the location manager may need one but cannot hire one unless the assistant is extremely inexpensive. These shows present an opportunity. By offering your services for low wages, you can work on a show, learn, and make contacts. To find low-budget films that are currently in production, read the industry trade magazines and newspapers (see below).

READ THE TRADES

The film industry trade magazines and newspapers (e.g., *The Hollywood Reporter, Daily Variety*) list those production companies that are starting up production. Contact these companies and find out if you can apply as an assistant (see above). In addition, apply as a scout (not as an assistant) to the very low or no-budget productions and student films. These productions are smaller and pay much less and therefore less expectation is placed on your performance. Very low (budgeted for under $1 million), no-budget (filmed on a "shoestring"), and student films do not expect to be able to hire the most qualified people for each job. A scout working on a no-budget or student film has an opportunity to make a contribution, learn about scouting, and do it with less pressure than on a budgeted production.

BUSINESS CARDS

Create business and Rolodex cards with your name and telephone number and the words "location scouting" on them. Use a list of production companies (from your local production directory, see Appendix A) to distribute the cards. You should enclose a letter of introduction that mentions your highly competitive scouting rate. Also, attend film industry trade shows and pass your cards out to everyone you meet. Upcoming trade shows are listed in the various industry trade magazines and newspapers.

GET LISTED

There are many film production resource directories (see Appendix A). Find those that cover your area and get your name listed under "location scouting." You may want to place an advertisement also. In addition, there are government film offices in most major cities that review filming applications and give permission to allow filming on their streets. Most of the time these offices know exactly who is filming what in their town. Contact your local office (see Appendix B) and ask to be added to its list of location scouts. Often a production company will contact the local film office to find a local scout. If the office does not publish an official list, drop off some of your business and Rolodex cards.

PRODUCTION ASSISTANT

As discussed in Chapter 10, the position of P.A. allows a person to learn and train for many other positions in production. Many location managers started as P.A.s.

Production Accounting

Career Profile

Job description
The production accounting department is responsible for bookkeeping, observing all legal and financial procedures, and estimating and reporting current and future costs.

Median income
Department head: $1600/week.
First assistant: $900/week.
Clerical assistant: $650/week.

Work closely with
All department heads and especially the producer and production manager.

Basic requirements
Bookkeeping and mathematical skills, computer expertise, and managerial personality.

Employment period
Preproduction, photography, and part of postproduction.

Duties

The production accounting department is responsible for keeping track of the thousands or millions of dollars that are expended by the production, estimating the finishing costs of the project during the shoot, and managing the crew's financial responsibilities.

During preproduction the production accountant assists the production manager and producer budget the production. The accountant is expected to be very familiar with all the costs involved in filming. If the cost of making the picture is more than the amount budgeted, the production manager and producer may turn to the accountant and say, "If we cut this scene and go to practical locations instead of a stage, how much will we save?"

When principal photography begins, the accounting department is responsible for the production's bookkeeping, for maintaining accounts payable and receivable, for payroll, for observing all legal and financial procedures, and for estimating, at each stage of production, how much it will cost to complete the shoot. During the course of a production, the producer and the financier of the film (e.g., studio, limited partnership, individual) will want to know exactly how much more money it will cost to finish production. Because the amounts of money, especially on a feature film, can be so large, the accountant's estimating accuracy is important. A film that is financially out of control can quickly go hundreds of thousands or millions of dollars over budget.

During postproduction, the accounting department will stay on to close the production books and may be responsible for overseeing the expenses involved in editing.

Nearly every feature-length production has a separate accounting department. Smaller productions such as music videos and commercials often use the production company's in-house accountant. Since bookkeeping requirements reflect the size and duration of a production, a music video or commercial company is able to use one or two accountants to handle all the company's projects. However, a feature-length production may film over the course of many weeks or months with a large crew and require its own full-time accounting staff to travel with the production. A mobile accounting department on location allows the flow of paperwork (i.e., invoices and checks) that is sent between the production office and the set to be processed without delay.

The head of the accounting department is called the production accountant (or auditor or controller), and her responsibilities depend on the size of the picture and whether the production is a studio or an independent

SAMPLE BUDGET (Low Budget Production)		
Title "The Shell"	Production company	

Above-the-line			Totals
	Accounting numbers	Description	
	100	Screenplay	
	200	Producer	
	300	Director	
	400	Cast	
			Subtotal
Below-the-line			
	500	Production staff salaries	
	600	Production operating staff	
	700	Extras	
	800	Sets	
	900	Props	
	1000	Costumes	
	1100	Makeup/hairdressing	
	1200	Sound/music	
	1300	Location/studio rental	
	1400	Tests/retakes	
	1500	Laboratory/film	
	1600	Insurance/taxes	
	1700	Editing/postproduction	
	1800	Overhead/miscellaneous	
			Subtotal
		Contingency @ 15%	

Figure 12 *Sample budget used in a low-budget feature. Production account-ants are very involved in creating a project's budget. These forms reflect how complex production budgeting can be.*

project. Typically, on an independent (nonstudio) film with a budget under $10 million, the accounting department is composed of the controller, a first assistant, and a second assistant. The controller works with all the department heads, analyzes costs, does spread sheet analyses of various contingencies, estimates or "controls" future costs, and reports to the pro-duction manager and the producer. Also, as the financial manager of the

			SAMPLE BUDGET DETAIL		
			(Low Budget Production)		
100	Screenplay		Detail	Amounts	Subtotal
		101	Story rights		
		102	Research/travel		
		103	Writer/screenplay		
		104	WGA reg.		
200	Producer				
		201	Producer		
		202	Executive producer		
		203	Associate producer		
300	Director				
		301	Director		
		302	Assistant		
		303	Secretary		
400	Cast				
		401	Lead players		
		402	Supporting players		
		403	Stuntpersons		
500	Production staff				
		501	Production manager		
		502	First assistant director		
		503	Second assistant director		
		504	Script supervisor		
600	Production/ operating staff				
		601	Director of photography		
		602	Camera operator		
		603	First assistant cinematographer		
		604	Sound mixer		
		605	Boom operator		
		606	Gaffer		
		607	Best boy		
		608	Electricians		
		609	Key grip		
		610	Best boy		
		611	Grips		
		612	Still photographer		
		613	FX person		
		614	Production assistants		
		615	Guards		
700	Extras				
		701	Extras		
		702	Stand-ins		
		703	Stuntpersons		

Figure 13 *Detail of sample budget for a low-budget feature.*

800	Sets			Detail	Amounts	Subtotal
		801		Art director		
		802		Set decorator		
		803		Set dresser		
		804		Lead person		
		805		Construction crew		
		806		Construction costs		
		807		Purchases		
		808		Rentals		
900	Props					
		901		Property master		
		902		Props purchase		
		903		Props rental		
		904		Prop truck		
1000	Costumes					
		1001		Costumer		
		1002		Wardrobe purchase		
		1003		Wardrobe rental		
		1004		Wardrobe miscellaneous		
1100	Makeup/hairdressing					
		1101		Makeup artist		
		1102		Hairdresser		
		1103		Supplies purchase		
		1104		Supplies rental		
1200	Sound music					
		1201		Sound transfer		
		1202		Dialogue editing		
		1203		Looping		
		1204		Sound effects		
		1205		Composer		
		1206		Conductor		
		1207		Musicians/singer		
		1208		Recording facility		
		1209		Music rights		
1300	Location studio rental					
		1301		Location manager		
		1302		Location rentals		
		1303		Permits		
		1304		Police/firefighters		
		1305		Studio rentals		
		1306		Studio personnel		
		1307		Dressing rooms		

Figure 13   *(continued)*

1400	Tests/retakes			Detail	Amounts	Subtotal
		1401		Makeup tests		
		1402		Retakes		
1500	Laboratory/film					
		1501		Negative stock		
		1502		Process and print dailies		
		1503		Black and white dupes		
		1504		Stock footage		
		1505		Answer print		
1600	Insurance/taxes					
		1601		Errors and omissions		
		1602		Workman's compensation		
		1603		Cast insurance		
		1604		Taxes		
1700	Editing/postproduction					
		1701		Editor		
		1702		Assistant editor		
		1703		Equipt./facility rental		
		1704		Coding		
		1705		Titles and optics		
		1706		Sound mix		
		1707		Preview screenings		
		1708		Supplies purchase		
1800	Overhead/miscellaneous					
		1801		Animals		
		1802		Catering		
		1803		Shipping		
		1804		Telephone		
		1805		Business license		
		1806		Accounting		
		1807		Legal		
		1808		Office		
		1809		Postage and supplies		
				Subtotal		
				Contingency @ 15%		
				Grand total		

Figure 13 *(continued)*

production, the accountant is expected to make sure the crew observes all the production's legal and financial procedures. Purchase orders must be properly filled out, check requests should have social security numbers, deal memos must be kept current, and so on. Although studio pictures tend to be larger and therefore the accounting department may also be larger, the responsibilities of the controller's staff are nearly identical.

The first assistant is responsible for doing the day-to-day paperwork, opening invoices, entering time cards into the computer, and generating checks. The "first" works very closely with the controller, and this is an excellent opportunity for him to train to become a controller by learning about the various aspects and responsibilities of production accounting.

The second assistant is more of an entry-level accounting position and is usually responsible for the clerical work and helping out wherever help is needed. On large shows (especially studio productions) the accounting staff may be composed of five or more individuals, and assistants may be given one specific area of responsibility such as accounts payable.

The major difference between the personality of the controller and the controller's staff is managerial acumen.

While the staff is responsible for the day-to-day nuts and bolts of book-keeping, the controller must interact with many of the people on the crew and manage their financial responsibilities as they relate to the production. Good controllers are not just excellent bookkeepers but are also very effective managers.

Because the accountant is well versed in the financial aspects of production and interacts with the individuals financing the film, production accounting often leads to production managing or producing.

Interviews With Production Accountants

> *Susan Gelb has many years' experience as a production accountant and associate producer. Her credits include* The Osterman Weekend, Crimes of Passion, The Howling, Children of the Corn, Two Moon Junction, Moving, *and the television series* Hell Town. *Today, Ms. Gelb is the director of features estimating for Walt Disney Pictures.*

How did you get started as a production accountant?

I began as a general manager of an arts organization where we had to do monthly cost reports. I oversaw that, and then I became an assistant to a controller in a television commercial company. While I was doing that I began to see all the paperwork there, and I realized that what was true for film was true for theater, just different labels. Then, my boss at that time was offered a low-budget independent feature, and I heard her saying that she was not going to take it. I asked her if she could recommend me for the job. She set me up with the interview, and I got the job as the show's production accountant. I did my first show with no assistants and no computer, but I was very lucky because I was replacing somebody else and the actual system of the books and the bookkeeping system was already set up. So I went in and parodied it until I understood what I was doing. Since I had been the general manager of a major arts organization, I knew how to manage and run a full set of books, but I did not know the words. I did not know what a gaffer was; I did not know what a grip was.

Are the general principles of accounting all the same?

Yes, accounting is all logic.

How would you describe the responsibilities of the production accountant?

You have a responsibility to the people who invest the money that all accounting and legal procedures are completely accurate and correct and that at any given moment the production is kept apprised of the current costs of the project and how much it is anticipated it will cost to complete. The first aspect is the nuts and bolts of bookkeeping, and the other is making sure that everyone knows how much the project is costing.

In addition to accounting skills, what special skills should someone possess to make a good production accountant?

If you are aiming to be an accountant, if you are not aiming to be an assistant but if you are aiming to go to the top, it is important to understand that production accountants are not the same as regular accountants. A regular accountant is someone who just sits down and adds up figures. Production accountants are much more managerial. They manage the crew. They have to handle people. The crew does not want to sign deal memos. The crew does not want to do purchase orders. The crew does not want to have to answer for every cent in its petty cash outlay. So to be a production auditor, you must have what would be considered managerial skills. You must be able to deal with people. And you must be able to interact with three of the more important people on the picture—the producer, the production manager, and the studio representative. You must be able to report to them and to make them feel comfortable that however many millions of dollars are entrusted in your care, it is being properly handled. You have to be able to please the people above you and at the same time manage the crew. The people who are going to break through and become the auditors are the people that have those skills or the potential to develop those skills. Another important attribute is not minding hard work because it is one of those departments that never stops working.

In terms of a career path, what other production positions can an accountant move into?

Production accountants become unit production managers (U.P.M.s) and producers because they know the whole business side and because sometimes a production accountant may do some of the work of a U.P.M. on an independent film. And the reason it can lead to producing is because you are often dealing with the people who have the money. So if you prove to them that you can run the right ship, maybe they will let you produce something of theirs.

So being able to run the right ship does not just mean keeping the books but also means estimating future costs?

Because of the big bucks involved, to lose control of a film can mean millions. Estimating is the newest trend because there is so much money involved. That is where your strength lies. The other stuff is automatic. Of course you keep good books. Of course the records are kept legally. Of course all that happens. But what does not of course happen is someone who knows exactly where they are today, not how much they've spent today but how much they've committed to today and how much they will have spent at the end of the picture. That is the skill of the people who are really good.

Paul Steinke is a production accountant for Walt Disney Pictures. His credits include Ruthless People, Big Business, On the Run, *and* Dick Tracy.

How would you describe the position of production accountant in a studio?

At the studio there is an accounting department and there are also the features and television accounting departments—two separate areas. The accounting department for the studio is strictly concerned with studio operations, and it is completely separate from the features or television estimating departments. These are the departments you want to get into if you want to be a production accountant.

The job of production accountant probably is not what an outsider would think it is. They think it is drudgery and just keeping books, and although it does incorporate that, it goes far beyond. In feature films especially, it is very hands-on and very involved with production. You have the same problems as a regular accountant—accounts payable, accounts receivable, payroll. It is like setting up a large company. In the case of a decent-sized film, you are going to have $10 or $20 million run through your hands in six months. It's a very large responsibility.

How are the various responsibilities divided in your department?

My assistants do most of the clerical work, which I oversee. This enables them to build a foundation of understanding production accounting. I have to deal with all the department heads, watch all the costs, and make sure that they are keeping me up to date. My primary function is to report back to the studio and to the producer and production manager on a daily basis. In any given moment I should be able to say where we stand financially.

What special abilities should a production accountant possess that a regular accountant does not need?

A certified public accountant will have great skills. He will understand debits and credits and journal entries—in short, basic accounting principles, but he will not know the running and timing of a production office or understand "production." There are so many things involved with that, that only ex-perience in production can teach you: choosing a location that best suits the film and budget, knowing what to estimate for building a hotel lobby, or knowing how much it would cost to dress a particular character.

How did you get started?

I was always good in math, and I love the film industry. I just was a movie fanatic. When I went to college, I was an international finance major. I thought accounting was too dry. I took business classes and a few film classes. When I graduated, I wanted to break into the film business, but I thought it was impossible. I flooded the market with resumes, all of them addressed to the top people of each company. Somehow one got to the right person at CBS, and it came down to the executive level of personnel, and they brought me in for an interview. So I started off at a part-time job at CBS as a financial analyst. I really did not know what was out there in the film business. My biggest regret was that I did not know about the position of production accountant when I was younger. If you are talented, you can make a good salary at a very young age.

What is the role of the production accountant during preproduction?

Preproduction is usually the most creative. You are given a script and a certain amount to budget. While the director is developing his ideas for the film, the producer and production manager come over to my side and we start developing a budget. Sometimes the budget will have been created by the producer or production manager way in advance, and sometimes a studio will tell me that the budget of a film is going to be $10 million and then we start creating that $10 million. The production manager has been hard at work doing the board, and we will start backing into that number and seeing if we can make the movie for that amount. If we figure it is going to cost $15 million instead of $10 million, that is where we begin talking about changes. For example, the way you get it to $10 million is to cut this scene, go to practical locations, not build, and so on. We look at the kind of film we have at the $15 million figure and the kind of film we can have if we decrease the budget a certain amount.

How can someone prepare himself to get into this area?

To get into this area you do not necessarily need a college education, but you have to be an intelligent person; you have to have street smarts. Education should not stop a person from pursuing it. You can start at a very early age, which, in fact, you can do with most production positions. So get into it right away if you know it is what you want, because you can excel.

Getting Started

READ THE TRADES

Read the industry trade magazines and newspapers to find a small or low-budget film that is starting up production. These productions usually have only one production accountant on staff, and this is a good opportunity to offer your services as an assistant. Perhaps the production accountant needs help but there is no money in the budget for such an assistant. If you step in at this time and offer your services inexpensively or for free, you have a good chance of getting started. This same technique would work if there is already one assistant in the accounting department but this assistant is overworked and can use help. (For a list of motion picture/television industry trade newspapers and magazines, see Appendix C.)

CONTACT STUDIOS

Sometimes the larger studios and production companies hire individuals to fill positions in its large accounting department. This department is not involved specifically with any production but is responsible for the corporate bookkeeping of the studio. Although these positions may only be for entry-level clerical help, any accounting-related job in a major studio brings excellent credibility and can be an effective first step. Contact studios and production companies in your area and fill out job applications even if no positions are open now. Big companies do not throw any piece of paper away.

TEMPORARY HELP

Sometimes when a large production is in full swing, the accounting department will suddenly find itself short of staff and may rely on a temporary employment agency to provide the extra help. There are many stories of "temp help" being kept on full time by the production company because the individual impressed the controller with her ability and motivation. Contact temporary help agencies and join those who are contacted by production companies to fill accounting related jobs.

OTHER PRODUCTION JOBS

Before becoming controllers, many production accountants worked in some other capacity in production. A number of them began as production assistants and production secretaries. These positions give individuals the opportunity to help a controller and work closely with the accounting department. By providing generous assistance and "hitting it off" with a controller, you may be invited to work in the accounting department of the accountant's next production. To get a job as a P.A., see Chapter 7. To become a production secretary, offer your secretarial skills to independent production companies for a very competitive salary.

CONTACT PRODUCTION ACCOUNTANTS

Although many production accountants already have a list of assistants, there is always time when there is a "crunch"—perhaps many productions are starting up at the same time, and assistants are in demand but in short supply. Call it timing; call it luck. If you can contact an accountant at this time and can impress him with your ability and your willingness to work for less, you may be called in immediately. But even if they have nothing for you now, send out letters and business cards with your telephone number and let them know who you are and that you are available for any work. They will keep your name and telephone number, and you may get a call next time they are in a "crunch." Production accountants are listed in production directories (see Appendix A).

Camera Department

Career Profile

Job description The camera department is responsible for photographing the best possible images for the production.

Median income Camera operator: $400 + /day.
First assistant: $300 + /day.
Second assistant: $270/day.
Loader: $230/day.

Work closely with Members of the camera department work with each other and with other technical departments (i.e., grip and electrical).

Basic requirements Technical knowledge, responsible, accurate, organized.

Employment period Director of photography: Preproduction and principal photography.
Members of camera department: One or several days before photography begins and during photography.

Duties

The camera department is responsible for photographing the finest images possible for the film or video production. The camera crew consists of the director of photography (also called D.P., cameraman, or cinematographer), the camera operator (also called second cameraman), the first and second assistants, and possibly a film loader.

The director of photography is the head of the camera department and works very closely with the director. The D.P. is responsible for achieving the photographic "look" that the director envisions. The D.P. communicates with the other technical departments (i.e., grip and electrical) and supervises all aspects of the photography, including lighting, camera movement, framing, and focus.

During preproduction the D.P. selects the camera, film, and lighting equipment needed and hires the key people of her support crew (i.e., gaffer, key grip, camera operator). During photography the D.P. is responsible for providing the director with the kinds of photographic images that he desires. Based on the director's input, the D.P. creates the look of the lighting, frames and designs the shots, and oversees everything that has to do with the photography. Directors of photography are usually chosen by the director based on their reputation for a particular lighting or framing style.

The camera operator is usually chosen by the D.P. and is responsible for actually operating the camera. The operator is in charge of keeping the action in frame, maintaining the composition established by the D.P., and photographing only the designated images. If the corner of a microphone appears or the shadow of a piece of equipment, then it is the operator's responsibility to see it and to keep it out of frame or, if that is impossible, to bring it to the D.P.'s attention.

The camera operator has several assistants—the first assistant, second assistant, and loader. The first assistant is responsible for the camera itself, that is, for making sure it is operating correctly, that it is clean, and that it has all the accessories required for the shoot. Before shooting, the first assistant cleans all the camera elements, installs the appropriate lens and filters, marks where the actors will stand, measures the distance between the lens and the objects being photographed, and establishes the correct focus and lens stop. While shooting is in progress, the first assistant maintains (or "racks") focus as the object being photographed moves closer or further from the camera. The job of first assistant carries with it a great deal of responsibility. If a single take is perfect—action, framing, lighting, and acting all flawless—and later the film is found to be torn from incorrect threading

Figure 14 *Productions can use any of a wide variety of camera systems. Cameras can be mounted on simple tripods like this one, hand held, or maneuvered by sophisticated cranes. Achieving the best shot is a precise and demanding task, requiring the crew to balance many different needs.*

or dirty from foreign matter in the camera's gate (opening where film is held in the camera), the first assistant is held responsible.

The second assistant cameraman (or second A.C.) is responsible for loading and unloading the magazines from the camera, preparing the slate for each take by filling in the correct scene, roll, and take number, and slapping the clapsticks at the beginning or end of each take. The second assistant also assists the first assistant wherever needed and is responsible for all camera department paperwork such as filling out the camera reports (detailed accounts listing the scene number, the number of takes for each shot, the amount of film exposed, and what the director has said about each take such as "print" or "NG" [no good]).

Loaders are used in large productions or when more than one camera is being used. The loader assists the second A.C. by using a changing bag to load the film into the magazines.

Although many productions employ individuals representing all five of the camera positions described above, many do not. On nonunion commercials, the D.P. is often the director and also operates the camera. The first assistant may be the only assistant and therefore must take on the responsibilities of the second assistant and the loader (i.e., filling out the camera reports, loading and unloading magazines). When productions hire a D.P., operator, and first and second assistants but do not hire a loader, the second assistant does the loading.

The camera department routinely uses a variety of specialty rigs, mounts, and other equipment and accessories to photograph underwater, in the air, and on the ground. One piece of equipment called the steadi-cam was invented in the mid 1970s and has revolutionized the way films and videos look. The steadi-cam allows an operator to walk or run near the objects being photographed while maintaining a steady, smooth, and level frame. This has provided greater flexibility in tracking subjects that in the past required long lines of dolly track. Other important pieces of equipment include helicopter mounts that allow a camera to be attached to a helicopter to photograph smooth, vibration-free aerial shots and waterproof casings that allow cameras to be taken underwater.

Interviews with a Camera Assistant and Operator

Dan Adams has worked in the motion picture/television industry as a first assistant cameraman for over ten years. His credits include numerous commercials, documentaries, and features.

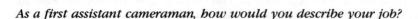

As a first assistant cameraman, how would you describe your job?

You are in charge of the camera. You make sure it works, it has all the accessories it needs, and it functions the way it is supposed to during filming. The day before the shoot you prepare the equipment. If there is anything wrong with the film the next day when they look at the dailies, they look at you. It must be in focus and it must not have any scratches.

How do you prepare the camera?

The camera comes in about 25 or 30 cases from the rental company. It means putting it all together. You check that everything works. You make sure that the power cables do not have a short and that the camera runs at the proper speed. Sometimes you shoot steady tests, and you screen the negatives to see if the camera is registering properly.

Is the job of loader the same as camera assistant?

In the United States there are usually two assistant cameramen on a feature or television series, called the first assistant and the second assistant. In England the first assistant is called the focus puller and the second assistant is called the clapper/loader. The second assistant puts marks on the floor, loads and downloads the film, and assists the first assistant. The first assistant is in charge of the camera, following focus and zooming with a zoom lens.

Where do you get the expertise to be a first assistant?

Through a process called apprenticeship. Where you work under other assistants, other cameramen. Most people have come up through the ranks; therefore, when you have a question, you can ask people how to do it.

> *Bruce Alan Greene has been a camera operator and steadi-cam operator on features, documentaries, and many other productions. His credits include* Patty Hearst, Time of Destiny, *and* Upworld.

Who hires the camera operator?

Generally, the director of photography wants a say in choosing the operator. When you have more than one operator for multiple cameras, stunts, or action scenes, sometimes the D.P. will not have any say and the production manager will just bring people in. But with one operator, usually the D.P. decides.

How would you describe the job of the camera operator?

The person who aims the camera and takes instructions from the director of photography that have been coordinated with the director. Some D.P.s will be very specific (about framing); others, if they feel comfortable with you, will just say give me a "two-shot" and they will not even look at it. Another big part of the job is making sure you photograph the movie and nothing that is not supposed to be there like a microphone or light. Sometimes you have to check the direction an actor is looking. Was he looking to camera right or left? This is important so that the scene will match when it is cut in (to the point-of-view scene of what the actor is looking at).

How does the steadi-cam keep the camera perfectly balanced?

First imagine a board balanced on the tip of a nail. The nail sits in a little hole at the board's center of gravity. Now the board is balanced on the tip of the nail so it will stay level. Imagine if you kept your hands off the board and kept the nail in the little hole. Now you could wiggle the nail around and change the angle, but as long as the tip stayed in the little hole in the board's center of gravity, the board would stay level. The movement of the nail is isolated from the board. The board, of course, is the movie camera. The other component is the spring arm that supports the weight and absorbs the shock. What makes it a real challenge and a specialty is that if you touch the board (camera) with too much force, it will tilt in every direction uncontrollably.

How much does the entire outfit weight?

Fifty to seventy pounds with the camera.

Does the assistants' role differ with a steadi-cam as opposed to a camera mounted on a tripod or dolly?

It is particularly difficult for the assistants because the distance between the camera and the actors is always changing, because with a steadi-cam you're always moving. It is more difficult for hitting marks and for focusing. The assistant must visually judge distances without reference marks to adjust the focus, and that is more difficult. On a dolly the camera moves on the track and you can measure the exact distance between a point on the track and a point on the set. But with a steadi-cam it is as if the track is wandering all over the place.

How do assistants adjust the focus on the camera?

They must keep their hands off the camera so they will not upset the balance. The way they adjust the focus is by radio remote control. There are little motors on the lens that spin the focus gear around, and the assistant turns a knob on a remote control box.

Is the steadi-cam used more in production each year?

Yes, because it offers the director and D.P. the freedom to move the camera without the restrictions of dolly and track and the time needed to set them up. As the number of skilled steadi-cam operators has increased, so has the acceptance of the steadi-cam in the film industry.

Getting Started

WORK IN CAMERA RENTAL HOUSES

The camera department is one of the more technically oriented departments in productions. Even the most basic entry-level job, the "loader," requires some technical expertise. One of the best ways to achieve that experience is to work in a camera rental house that services motion picture production. At a rental house you will learn how the various cameras and equipment work, how to assemble and troubleshoot them, and how different cameras, lenses, filters, mounts, and accessories are chosen for various applications. In addition, working at a rental house will give you an opportunity to meet camera assistants and to impress them with your technical knowledge and offer your services. Remember that personality is as important as technical know-how. First A.C.s look for a second assistant or loader who is easy to work with as well as knowledgeable.

CREATE A SAMPLE REEL

It is possible to get an assistant's or operator's job or even a job as cinematographer (on a very small production) by showing the producer, director, and production manager a sample reel of your work. A sample reel may consist of a single finished film or video that you have shot from beginning to end or a series of sequences from various productions that you have shot and that are edited together. The reel should exemplify your cinematographic skills, abilities, and talents. If a low-budget production feels that your reel is consistent with the degree and complexity of challenges that you will be facing on their production, then you may be hired.

Technical knowledge is usually also important. Many low-budget productions already have a director (the director may be producing the show), but often what the director looks for in an operator is someone who knows the equipment, someone who can fix it if it breaks and maintain it so the best images possible will be photographed at all times. With technical know-how and a reel you can approach low-budget productions (i.e., documentary, educational and industrial companies) and ask if you can make a contribution. Remember also to keep your salary expectations low in the beginning.

READ THE TRADES

If you live near major production centers, read the trades to find out if there are any low-budget productions (i.e., industrials, educational or student films) that are crewing up. Contact these productions and volunteer your services. If you have technical knowledge, you may land a technical job (i.e., assistant cameraman). If not, you may be brought on as "utility," someone to help carry the camera equipment, set up the tripod, and so on. If you are enthusiastic, motivated, willing to keep your salary expectations extremely low, and willing to participate in the smallest and most low-budget productions, you should be able to get a job in the camera department.

GET LISTED

Once you begin to accumulate experience on the set and technical know-how about the camera, get your name listed in the various production directories in your area. Most directories have listings for assistants and operators. Make sure you have an answering machine or service connected to the telephone number that you list, and check your messages frequently.

CONTACT ASSISTANTS AND OPERATORS

With some experience and technical knowledge, you can contact first assistant cameramen and operators and make your services available to them. Because of the very important role played by all members of the camera department, most experienced operators and assistants tend to work with people whom they know they can count on. It can be difficult to gain someone's trust in this fashion, but if you volunteer to assist them on a couple of jobs and can slowly win their trust (i.e., first by carrying equipment and not breaking it, then by changing lenses and loading film in the magazine), you can begin to nurture working relationships with experienced professionals.

ATTEND FILM SEMINARS

Film and video seminars are usually announced in the industry trade magazines (see Appendix C) and by colleges and universities with film departments. By attending these workshops or seminars you can learn skills and techniques from the experts and meet others in the motion picture/television industry. Keep an eye out for production-oriented seminars or, more specifically, any workshop or seminar that will focus on the photographic arts.

Art Department

Career Profile

Job description Responsible for translating director's ideas into tangible visual images.

Median income Salaries range depending on position.
Head of art department: $300 + /day.
Entry-level assistant: $125 + /day.

Work closely with Art department personnel work with each other. The department head works with many of the other departments.

Basic requirements Entry-level jobs require an artistic sensibility, initiative, and the ability to understand and take direction.

Employment period Flexible. Can be preproduction, photography, and postproduction (wrap).

Duties

The art department is responsible for translating the director's vision into reality on the set. The way a set looks, the furniture, design, colors, fabric, textures, artwork, and decor all come under the responsibility of the art department. To discuss how the art department translates and amplifies the director's ideas and visions, it is useful to examine the various jobs or components of the department.

The head of the art department is called the production designer. The production designer, a relatively new position in the film industry, is responsible for creating the overall "look" of the film or video and for hiring the individual members of the art department. In addition to supervising the art department, the production designer works closely with the other departments (i.e., makeup, costume) to oversee the entire visual presentation.

During preproduction the production designer meets with the director to discuss the film's overall texture, mood, and visual design. For example, the director might explain that the film's plot (e.g., corporate extortion) would be enhanced with an ultramodern look using hi-tech sets of glass and steel, art furniture, black and white paintings, cold colors, and expensive glass decor.

The production designer would create the overall look of the film, where each location and scene fits like a jigsaw piece into a master design. The production designer would communicate the ideas with the art director and other individuals in the art department. In addition, the production designer may discuss the "look" with the key people of other departments, explaining to the makeup artist the importance of using cold bright colors and emphasizing to the costume designer the role of ultramodern chic clothes to enhance the high-tech physical environment.

Based on the information supplied by the production designer, the art director designs and decorates the sets. The art director is the "architect" of the production and will create sketches, plans, and models of the various sets. This position requires both a sense of balance and design and a thorough knowledge of architecture. On some productions, the art director will have the help of a set designer, who will plan the construction of the sets based on the art director's sketches and models.

Enhancing the work of the art director is the set decorator. The set decorator is the "interior decorator" of the production and is responsible for selecting and placing all set dressing needed on the set. Set dressing is any item (furniture, artwork, decor) that is not movable or portable and is not mentioned specifically in the script. For example, if a script reads:

The senator enters the study and walks to his desk. He opens the desk drawer and takes out an old hardcover book. When he opens the book the audience realizes that it's not a book at all, but a box built into a book. Inside the box is a diamond. He pockets the diamond, closes the book, and walks downstairs.

The set decorator is directed to design this set traditionally. With creative supervision from the production designer, the set decorator brings in and arranges a roll-top desk, dark birch shelves filled with books, a fat leather chair and couch, thick shag carpeting, and the stuffed head of a black bear placed on the wall.

While the set decorator is responsible for the set dressing (all the furnishings and decor needed to decorate a set but not specifically mentioned in the script), the property master is responsible for "hand props" and those items mentioned in the script. "Hand props" can mean almost anything that is carried by an actor. Although a script might not mention that a character carries a briefcase, if he does, it is supplied by the prop master.

In the example above, the diamond is specifically mentioned in the script and is therefore considered a prop. In addition, the special book that is actually a box is also mentioned, and therefore the property master would be responsible for finding, securing, and placing onto the set both the diamond and the special book. Contrastingly, the books on the birch shelves that line the walls fall under the category of set dressing and are therefore the set decorator's responsibility. Obviously, there are occasions where the responsibilities of the set decorator and property master may overlap.

Most productions shoot at a variety of locations, and the production designer, art director, and set director may be at a future location planning and prepping (preparing the location) while the rest of the crew is shooting somewhere else. In those cases a set dresser assists the set decorator by maintaining the condition of the set during shooting. As camera angles change and furniture is moved, the set dresser must reposition the dressing, maintaining the look of the set as the set decorator intended.

The art department relies on a great deal of manual labor to pick up and deliver furniture from rental houses, decorate the set, hang paintings, and return furnishings. Decorating each new set is analogous to selling your house, packing up everything, moving, unpacking, and redecorating. As a result, most art departments rely on two other positions to supply much of the physical labor—leadman and swing gang. The leadman, as the title implies, leads by working ahead of the rest of the production. On Monday the leadman may be picking up and delivering furniture for the Wednesday location. The swing gang is the leadman's crew and do much of the heavy moving and delivery work. Both the lead man and swing gang may also help out the other individuals in the art department. They may assist the set decorator by putting pictures in picture frames, or they may help the set dresser by rearranging furniture on the set.

Some of the other positions in or associated with the art department include draftspeople (drawing sketches for sets, props, and costumes), model builders (designing, building, and operating all miniature models), prop makers (designing, building, and operating any special props), greensmen (selecting and maintaining all greenery), set construction (e.g., carpenters, painters, plasterers, welders) and the art department coordinator (coordinating art department activities, rentals, and returns).

Interviews with a Production Designer and Property Master

> *Michelle Minch has worked as a production designer on features, music videos, and national commercials.*

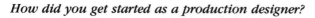

How did you get started as a production designer?

I studied veterinary medicine before I came to Los Angeles. I had no idea that there was such a job as production designer. I began in still photography, working as a photo stylist with an emphasis on props and set decorating. In still photography you are much more broadly based than you are in film. I did a little of everything—wardrobe, food styling, and makeup. I realized that I really liked set decorating, and so I began looking for work, assisting art directors as a set decorator. I worked for several art directors and realized that I could do their job as well as they could.

Is that how a lot of people move into positions of art director and production designer, by working their way up the ranks in the art department?

Yes, definitely. Almost any entry-level position is a good way. To get an entry-level position the person should be willing to work inexpensively or even for free for a little while (to make contacts and get credits for your resume).

What talents or aptitudes should a person possess to do well in the art department?

They have to be ready, willing, and able to do almost anything at a moment's notice. The higher the degree of intelligence, the better. Paying attention is very important. If they own tools or know how to use them, that is definitely to their advantage, or if they have specific skills such as being a painter and being able to act as a scenic painter, they will be hired before somebody who cannot do that. A good attitude is also important, such as somebody who keeps a stiff upper lip and is willing to work long hours for little pay.

How would you recommend someone get started?

The best way to get work in the motion picture industry is to work in video first. There is much more video production than film production. You have a better chance of getting in and making contacts that way.

How do you begin in videos?

Call production companies that are listed in production directories and offer yourself as a P.A. As you work a show as a P.A., meet the art director and meet the other people who are working on the crew. Try to make a good impression, and then ask everybody you can who they know that you can call to ask about doing other work. Be persistent!

Once you have broken in and gotten your first job, how do you continue to work?

By networking. But the best way to get your next job is to do a great job on the one you are working now, because if you do your job well you will always be noticed.

Dan Morski has been a property master on numerous feature films and national commercials. His credits include Weeds, Alligator, *and* The Last Resort.

▼

How did you get started as a property master?

I majored in broadcasting and film at college. One summer I came out to Los Angeles. I had lined up some work as a production assistant on a film through one of my classmates. When I got out here it was "What film?" So I was left standing in a phone booth at Alameda and Riverside in the heat of the summer, and I had nothing. But I knew I wanted to work in the business, and I had to start somewhere. I finally got a job as a runner, and I got into props by taking advantage of an opportunity. The commercial companies I had been working for hired property masters but not their assistants. There was not even an assistant prop category. So the runner or P.A. would assist the property master, and that is how I spent the majority of my time. There was one particular property master who would get jobs and say to me, "This is the job, Dan. This is what you have to do. I will check back with you later." Later, when the opportunity came along to be a property master, I was able to do it.

How would you describe the job or purpose of the property master?

To enhance the performance of an actor or actress in a given scene by providing implements and hand props that make the situation come alive. For example, who would Dirty Harry be without his 44 Magnum, or Darth Vador without his light sword? Indiana Jones without his bull whip? In commercials the property master is solely responsible for the product. We make the beer look appetizing with spritz and ice and the soda cans dance with a perfect pour.

What abilities and skills does a property master need?

You have to be patient and conscientious and must be able to anticipate what is coming up next. Over the course of a ten-week shoot you have to stay on top of it or it will bury you. It is also helpful to have a background in crafts, carpentry, mechanics, arts, and drawing and to have a good strong back. There is a lot of physical work to it as well.

Are most prop masters men?

I have worked with a lot of women whom I could have beat arm wrestling, but that does not stop them from doing an excellent job.

What do you look for in an assistant?

I look for somebody who is resourceful, who can pull things out of a straw hat. I look for somebody who has some mechanical ability, who can handle tools. I also look for someone who is conscientious and who has a good attitude and who can maintain the same state of mind from the early morning until the end of the day.

With all the levels within the art department, the art director, the set decorator, and so on, how do you determine exactly what areas you are responsible for?

Coordination takes place between the art director, set decorator, and prop master. For example, the actor comes onto the set, which is an office, and sits down at a desk and opens the drawer. The desk itself and the contents of the drawer would be the set decorator's responsibility. If, however, the actor were to reach into the drawer and pull something out, a makeup compact or a wallet, that would be something that the propmaster would have been responsible for placing inside.

As audiences get more sophisticated, does it require greater sophistication on the part of the property master to create even more amazing props?

Yes. It is getting more specialized today. There is more rigging and more specialty kinds of props. Recently, I had to make a soft drink come out of a can in a wave. I had to turn the can upside down, turn the camera on its side, cut the top and bottom out of the can, put an apparatus inside the can, attach it to a servomotor arm, and attach the arm to a tube that would undulate from side to side and make the product come out in a wave.

How do you figure these things out?

That is what makes it so interesting to me. It is a challenge. You sit down and think about it, draw on past experiences, and come up with an idea. There is a joke in my department—that I will go and refer to the book on that. There is no book.

Getting Started

The jobs of swing gang, leadman, and assistant to the set dresser or property master present opportunities for individuals to begin working in the art department. Starting in the art department is similar to starting off in any other department in film or video production. Your first job will be the most challenging to obtain. After that you will have begun to accumulate contacts and experience on the set. By staying in touch with the people you meet on the set, your lifeline to new work, your future jobs will become easier to secure.

SWING GANG

In addition to their responsibilities of moving and delivering furnishings, the swing gang is responsible for providing general assistance to other individuals in the art department. Members of the swing gang may be called on to do a variety of things, including driving the art truck (on non-Teamster productions), laying down carpets, and helping to arrange exterior sets. To get a job on the swing gang, contact production companies, art directors, and production designers. Use the trade magazines and newspapers (see Appendix C) to find companies currently in production and who are hiring. In addition, use production directories (see Appendix A) to find complete listings of all production companies. Send out letters and business cards and make telephone calls. Most people in this business maintain a list of art "utility" people whom they can call.

READ THE TRADES

When you are beginning, you have a better change of getting hired by a student film, low-budget independent feature, or industrial or educational video. Because these smaller productions do not have the money to attract experienced professionals, they rely on individuals with talent and motivation who are willing to work for less. Depending on your previous art experience, these productions may hire you for nearly any position from art director to prop master's assistant. Student films, small independent features, and instructional videos that are beginning production are listed in the various motion picture/television trade publications (see Appendix C). Contact these productions and ask for the production manager, production coordinator, or the art department. Meet with these people if at all possible. If there are no openings on this production, there may be one on the next project.

GET LISTED

Some production directories have categories for art department "utility" personnel. Get your name listed and perhaps pay for an advertisement. Have an answering machine or service connected to the telephone number that you list, monitor your calls daily, and return work-related calls immediately. Production directories are used by production companies as a dependable source of equipment and personnel information all year round.

CONTACT ART DIRECTORS, PRODUCTION DESIGNERS, SET DECORATORS, SET DRESSERS, AND PROPERTY MASTERS

Experienced art department personnel are always looking for people they can count on to do a job and who can make them look good. A good assistant (swing gang, assistant property master) is someone who can play a dependable supporting role, who is a good problem solver, and who can take the initiative when required to do so. If a company stops shooting because the director suddenly needs flowers on the sets, a member of the swing gang may be sent on an "emergency run." Perhaps it is 9 PM and the set dresser sends you to a particular store to get the flowers. You get there fast but the store is closed. You cannot return to the set emptyhanded. You take the initiative, find a telephone, and call every flower store and nursery within ten miles. You find one that is just closing as you call and you convince them to stay open for you. Later, when you return to the set with the flowers, you look good, the set dresser looks good, and the company has not lost any time. This is the only way to act in production. You simply cannot return to the set without the flowers and say to the set dresser, "Sorry, the store is closed. What do you want me to do?"

People who work for the art departments of films and video productions are resourceful individuals who are good at responding to the sudden changes that are always a part of working in the art department. These professionals are always on the lookout for people who, like them, are handy, helpful, and resourceful. Convince prospective employers that you have these qualities, and when you are working, prove it to them.

Use production directories to find art directors, production designers, and other art department personnel. Send out letters and business cards and make telephone calls. Most people in this business maintain a list of people whom they can rely on. To get a job on the swing gang, promote yourself as an individual with an artistic eye, a strong back, and a willingness to take direction.

START AS A PRODUCTION ASSISTANT

Many people working in the art department of film and video productions began their careers working as a P.A. As a P.A., an individual can learn about the various positions in the art department firsthand, accumulate time on the set, and get paid. A P.A. who exhibits a particular affinity for the art department may find it easy to move into that department in some capacity. To learn about the job of production assistant, refer to Chapter 7.

Support Services

For nearly every person working in this industry, there is a different technique for beginning a career in film. It is clear that you must be creative to get started in the motion picture/television industry. It is, therefore, important to be aware of other avenues and possibilities. In addition to the numerous entry-level positions in production that offer the film student a chance to get started, there are many support services that provide similar opportunities.

There is a natural tendency for people to hire those people whom they know or have met. As a result, many individuals began their careers by working in or starting their own support service, where they became known to many individuals involved in the production of films, commercials, television, and videos.

While fully staffed production companies appear to be able to produce singlehandedly a film or video, most productions rely on a number of support services to either enhance the look of their film, save money or time, or furnish services that the production company simply cannot provide.

The term *support service* has a wide range of definitions, but generally it means those businesses that provide an important and sometimes critical service to a production company that the company does not or cannot provide for itself. Most support services are relatively small operations that were created by entrepreneurs. These services include caterers, prop houses, janitorial/strike services, animal trainers, casting directors, choreographers, location services and libraries, payroll services, camera cars, picture cars, motorhome rentals, and security guards.

An even broader definition of *support service* includes the larger technically oriented businesses such as film processing laboratories, film libraries, optical services, postproduction facilities, and special effects.

In this chapter we will discuss the far more numerous smaller operations that provide the individual with the opportunity to work closely with the motion picture/television industry and to learn about the industry from the inside. Many of these businesses can be started with little capital and can provide an effective springboard for getting started in this industry.

Location Services

Clients include:
Location scouts/managers
Production assistants
Production managers
Art directors
Production designers (low-budget production)
Directors (low-budget production)
Producers (low-budget production)

Today, more productions shoot at real locations than ever before. These real, or "practical," locations include everything outside a studio or backlot. A real house, restaurant, store, estate, and apartment building are all examples of practical locations. Location scouts and location managers are responsible for finding and securing practical locations, but occasionally, because of a lack of time, they must take a shortcut. Enter the location service. Location services are companies that represent properties as an agent would an actor. These services provide selections of locations that are available to filmmakers and act as middlemen between the property owner and the production company. Each property is photographed exteriorly and interiorly from many angles. The photographs are mounted in portfolios and are made available, to be viewed for free by the busy location scout or manager or another member of the production, including the producer or director on a low-budget show.

The pictures are borrowed from the library so the company has time to decide. When the director chooses where she wants to shoot, the location service will negotiate the best rental price it can for the property owner without loosing the deal. This benefits the vested interest of the service, which may keep 20% to 40% or more of the total location fee. In fact, most location services require the production company to pay them directly, and then they reimburse the property owner.

Although production companies use location services as a "last resort" (because they tend to inflate the cost of a location rental), there is a growing dependence on them. Also, most production companies are aware of the commission structure and are always on the lookout for a service that does not disproportionately inflate the cost of their properties.

The way to start a location service is rather simple. First you must find those properties that are in demand. The more properties you represent and the bigger and grander they are, the more frequently they will be rented. (Production companies have a natural attraction for the biggest, the best, and the most beautiful locations.)

Locations that fit these criteria include

- Warehouse space that production companies can use as a soundstage

- Mansions, estates, and large houses

- Any unusual location (e.g., a high-tech beach house, a cannery, an old-fashioned hotel, a large indoor swimming pool)

In Los Angeles, a production company will pay $2,000 or more a day for a warehouse space that can be used as a stage and pay $5000 a day for a particularly large and opulent estate.

Once found, each property must be signed up and photographed. Property owners should sign a document that stipulates that they agree to allow you to photograph the property, to show the pictures to production companies, and to act as the agent for the property during negotiations. (For a legal, binding document, have an attorney draw up the agreement but keep it simple.)

Once you have the properties signed and photographed, place the photographs in a report-size portfolio (i.e., between pages of plastic sheet protectors) and show the pictures to those individuals who are responsible for finding locations for filming—location scouts, location managers, production managers, and, on low-budget productions, production designers, art directors, producers, and directors. The marketing approach is critical. Mailers, brochures, and telephone calls are all effective. Advertise your service in the various production directories and use whatever means available to keep the pictures of the properties that you represent in front of the individuals who decide where to shoot.

Computerized Location Libraries

Clients include:
Location scouts/managers
Production assistants
Production managers
Art directors
Production accountants
Screenwriters
Production designers (low-budget production)
Directors (low-budget production)
Producers (low-budget production)

Recently, film liaison offices have been created all over the United States and abroad to attract filmmakers into their areas. These offices offer a one-stop source for local production information, including pictures of locations that are available for filming in their region. These government-sponsored film offices make their "location libraries" available free of charge and allow production companies to negotiate directly with the owner regarding the rental fee.

Today, new business called *computerized location services* have been created to compete with these government-sponsored film offices or to enhance access to their library for a fee. These new businesses use computers to collect, store, and retrieve specific locations using various location specifications.

Most government-sponsored film offices have files of photographs. To find the "right" house, someone from the production company may have to look through boxes of pictures. However, with a computerized location library, locations that meet certain physical and financial specifications can be found by simply punching a keyboard.

For example, perhaps the most important location in a script is a ranch house. The screenwriter wants to know if a ranch house exists with a sunken living room. If it does, he would like to write a special scene that takes advantage of this design. According to the schedule, the company will be at the "ranch house" for 15 days of shooting. The production accountant is interested in how much such a house will cost to rent. A rental fee of $1000 per day versus $4000 per day will make a large difference on the location budget. The art director must know what color the house is, and the location manager wants to know if there are any other ranch houses with sunken living rooms in case the first one is too expensive or is not approved by the director.

Here, the computerized location library is a great help. The computer will search its files for locations that meet these specifications (i.e., ranch house with sunken living room) and then print out or display a list of the addresses and telephone numbers of those properties. The approximate rental fee is also included. Each of the computer's selections has been previously photographed, interior and exterior, and the production company borrows the photographic file. The company pays the computerized location library for access to the computer and leaves with pictures, addresses, telephone numbers, and approximate cost of each filming location. Once a property is selected, the company negotiates the rental fee directly with the property owner.

Computerized location libraries are a brand-new industry. To start one, the best thing to do is to contact film liaison offices in your area (see Appendix B) and see if they would like their locations computerized. Then let location scouts and managers, production assistants, production managers, art directors, production accountants, screenwriters, directors, and producers know that you have location information for your area available at your fingertips.

Janitorial and Strike Services

Clients include:
Location managers
Production managers
Production assistants
Production coordinators
Set construction foremen

Although not the most glamorous business in the motion picture/television industry, janitorial and strike services are fairly inexpensive businesses to initiate and provide an opportunity to meet a number of production personnel.

These services are called in after filming has wrapped to return the location to its original condition through cleanup, maintenance, and simple repairs.

Most strike services will go out to the location and provide a free estimate. Charges usually begin at $150, with additional fees for cleaning drapes, vacuuming carpets, etc. To start a strike service, advertise in your local production directory and contact location managers, set construction companies, production managers, and production coordinators.

Property Rental Houses

Clients include:
Production designers
Art directors
Set decorators
Set dressers
Leadmen

Casino equipment, clocks, sculptures, aquariums, medical equipment, antiques, nautical brass, pinball games, old photographic equipment, billiard tables, jukeboxes, satellite dishes, and furniture are all items supplied to production companies by prop rental houses.

Technically, a prop is any movable item used on a set or specifically mentioned in the script, and set dressing is defined as furnishings, art work, and so on used to decorate the set. However, contrary to the definitions, the word *prop* in the term *prop house* can mean both set dressing and props.

Production companies rely on prop houses to supply the various items and furnishings needed on the set. Prop rental houses rent their items to production companies for fees based on the value and scarcity of the item and how many days it will be used.

There are three ways to get started in this business:

1. Provide the same items other prop houses provide at a lower cost.

2. Supply unique and hard to find props.

3. Do both.

A number of prop houses have built their businesses and reputations around their ability to procure and rent unusual props such as antiques, hi-tech equipment, and expensive artwork. Production companies are constantly trying to make their films and videos innovative, and they rely on new and innovative props to help them create that look.

Working in a prop house or starting your own is a good way to meet members of the art department of different productions. If you would like to create your own business, begin by contacting art directors in your area. Tell them that you plan to start a prop house, and ask them what kinds of props (and set dressing) they sometimes need but cannot find. (You can contact art directors by using the various production directories; see Appendix A.) Also, examine the prop house listings in your local production directories to evaluate the kinds of props currently available.

Camera Rental Houses

Clients include:
Directors of photography
Camera operators
First and second assistant cameramen
Loaders

Working at a camera rental house provides an excellent opportunity to learn about various photographic equipment and its applications and operations and to meet many individuals currently employed in the camera department of large and small productions.

Camera is considered a technical field for a good reason. Much of the equipment used in videotaping and filming is highly sophisticated. Accessories must be assembled in specific ways; certain lenses work with certain cameras while others do not; and cables, filters, battery packs, chargers, monitors, tripods, friction heads, mounts, and other pieces of equipment must be used and maintained properly.

Most of the people working in the camera departments of production companies are familiar with a wide variety of equipment. Working at a rental house provides an excellent learning environment to acquire this knowledge and to meet potential employers as you master the applications and proper uses of the various equipment and accessories.

Grip and Electrical Rental Houses

Clients include:
Gaffers
Key grips
Grips
Electricians

The grip and electrical departments in a production often depend on rental houses to provide a great deal of their equipment. As discussed in Chapter 10, working in a grip and electrical rental house provides an opportunity to become familiar with the names, applications, and purposes of various equipment and to meet key grips and gaffers who may need another person to "fill out" their crew.

Many grips and electricians began this way and moved directly from the rental house to a job in production.

Part IV

Producing and Directing

The Business of Producing

Working and developing a career in the motion picture/television industry requires a clear understanding of the purpose of most productions. For better or worse, the object for most commercial producers is to create a product that will make a profit. To do this some producers have relied on recreating already proved productions (e.g., sequels, "spinoffs"), while producers working with lower budgets have prided themselves on going against the trend to profit from initiating their own following. The success of films or videos is sometimes as unpredictable as a day at the races. Bet heavily on the favorite to make a decent profit, or bet a little on a longshot to make as much or more. Of course, whether you have a winner depends on which horse comes in.

Every production begins with a combination of the following elements: the property (script, book, play, song, or creative "concept"), a star or stars, a director, a distribution deal, and a source of financing (an individual, limited partnership, corporate sponsor, studio, or distribution company). The individual who is responsible for bringing together enough of these elements and enough of each element to begin production is called the producer.

Most good producers are good dealmakers. Production is an expensive endeavor, and the producer of a large budget production must convince a financier to spend millions of dollars. Smaller-budget producers may need thousands. There are hundreds of ways to make good deals. Many producers rely on having a good package (e.g., a famous star who wants to make the picture, a script by a successful screenwriter, a great director who wants to do the project). In the minds of prospective investors, such packages lessen the risk of a financial disaster. But films such as *Heaven's Gate* and *Ishtar* have proved that it does not work all the time.

Distribution is also important. Good producers know exactly how their film or video will be distributed once their production is completed and approximately how much money they and their investors will make. In fact, most producers try to presell the ancillary rights (i.e., foreign sales, video-cassette, cable television) before the production is completed.

There are as many kinds of producers as there are productions. Some producers work in the major studios (i.e. Warner Bros., Paramount, and Disney). Major studio producers oversee projects transferred to their care by studio executives. Other producers work independently. These independent producers start from "scratch" each time, finding financing, working out a distribution arrangement, and supervising the production in the making.

Some producers are very "hands on." These producers, sometimes called line producers, stay near the production office during preproduction and work closely with the director, the production managers, and the production accountant to make sure that the director is getting what she needs without going over budget. During photography the line producer is on the set answering questions, making decisions, and solving problems. Other producers are less "hands on" and stay in their office evaluating the production's progress from reports, budget analysis, and screening the dailies.

To become a profitmaking producer requires knowledge of many of the business and technical aspects of production and distribution including rights and royalties, insurance, and packaging. The many ways of getting this information include books, seminars, and working in production.

Like the other motion picture/television careers discussed in the previous sections, there is no single method to become a producer. Some producers, both big budget and small budget, have worked their way up the film industry ranks; others came in the side door with connections to potential financiers. For most producers, producing was a goal deemed possible only after they had acquired the information, paid their dues, and developed the contacts and the confidence needed to proceed.

Ronald G. Smith, production supervisor for Warner Brothers' Feature Films, helps to delineate further these introductory aspects of producing.

Interview with Ronald G. Smith

> *Ronald G. Smith is a studio producer who has spent most of his career involved in the various "on-line" aspects of motion picture production. His credits include executive in charge of production of* Scarecrow and Mrs. King *and associate producer of* Ghost Story. *Today, Mr. Smith is the production supervisor for Warner Brothers' Feature Films.*

What does a producer do?

The producer oversees the project. [In the case of a studio film] often the producer will bring the project to the studio. He makes sure that the film gets shot for the amount of money it is supposed to be shot for, that the creativity is there, and that the director is getting what he needs to sell the film and tell the story.

What is the difference between the line producer and the executive producer?

The line producer is a production-oriented producer; he knows the nuts and bolts of production. Executive producer is a title designated to someone who may have very little to do with the actual production.

The executive producer may be putting up the money.

Right.

What special skills or abilities should a producer have to be really good?

A producer should keep the actors and the director happy and make sure that the production gets done in the allotted time. Sometimes there are creative differences. The actor, director, and producer may interpret scenes in different ways. It is important for the producer to get everyone to sit down and to work out the differences. Filmmaking is a team effort. To me it is like being in sports. It takes 11 men to play football and 9 men to play baseball, and it takes the number that makes up the crew—whether it is 20, 50 or 100 people—to make a film work. Everyone works together. It is not one group who grandstands and gets all the glory. Everyone must work together for one common goal to get a good product, a film that you have a lot of good feelings about. I think a lot of that starts at the top and works its way down.

To continue the football metaphor, would the producer be the quarterback or the coach?

The producer is the coach, and the director is the quarterback. The production manager is like a lineman. He opens all the holes and gets no glory. The actors are the backs; they have to carry the ball metaphorically and give a good performance.

Where do producers come from? How do they get started?

A lot of times you can option a property from anywhere from $1000 to $100,000 and get the rights and then get a studio to back you. It does not take a lot of training, but it is better if you know what is going on. To me [the best way] is going through the ranks and working your way up that really makes you savvy and a good producer.

I began as a P.A., working on the set. I worked with [Steven] Spielberg on *Sugarland Express,* and I worked my way up the ranks, second A.D., first A.D., production manager, and producer. I have done all that. If you are going to be good it takes time, but whenever the opportunity arises you will be ready for it.

Can a producer choose a particular director to work on a film?

If it is a studio project, generally it is a collaboration with studio executives to determine who would be best suited for the job.

Do you think it is getting harder or easier to work in this industry?

I think right now is the time to really get going in this industry. When I first started in the early 1970s it was just starting to take off. New directors such as [Martin] Scorsese, [Francis Ford] Coppola, and [George] Lucas were just starting out. The industry is going through a period of growth now, and I hope that it continues for a long time. I think it is a great medium. There is no place else like it.

The Art of Directing

For many film students, the other jobs in production are simply a means to an end, with the end being, of course, to direct. Because directing is the goal for a large number of individuals and yet few people have the opportunity to direct early in their careers, the job of directing frequently becomes misconstrued and idealized. Yes, the director is the creative architect who is responsible for welding together the other elements in the production to form a cohesive narrative. But directing is more than being the creative author. It is an incredible challenge on many levels, including imagery, craftsmanship, and organization.

To be a director requires creative and technical skills and the ability to lead. Becoming a director and having the opportunity to direct has much more to do with possessing these abilities than with the notion of a "break." In other words, breaks come to those who already possess these qualities. Few directors in the history of cinema have been successful without being a creative visionary, having a technical knowledge of the medium, and having the ability to inspire others.

But what is most important to be a good director? Which ability takes precedence over the others? Which skills can be learned and how? Can a good leader create a great film without having a vision? Can technique alone make art?

Film and video production is still voyaging through its age of discovery. The motion picture/television industry is no more than eighty years old. It is an industry that has only begun its journey of exploration. Each year, the medium expands as new visions of the world are created, new techniques are invented, and new questions are asked. It is instructive to compare this age of discovery with another.

During the fifteenth and sixteenth centuries, sailing ships, two- and three-masted schooners and clippers, roamed the seas. Some ships followed the well-plotted sea lanes while others sailed beyond the edges of the charts into unknown waters. All ships, whether sailing for a visible port or to find a new continent, were composed of similar elements—the ship itself, a crew, a captain, charts, and a destination of some kind.

The ship was the medium for transporting people over the water. The crew provided the muscle to carry out the captain's orders and to make the ship point in the proper direction. The charts helped the captain plot the course, and the destination provided the goal to the voyage. Indeed, every ship had a captain whether it sailed in the sea lanes or left the lanes to sail in uncharted seas.

Metaphorically, productions are similar. Film and video is a medium for transporting people to another place or time. The crew provides the techniques, knowledge, and muscle to move the production forward. The script is similar to a chart by helping to plot the course. And the director, the captain of the production, decides where the production is going.

Some directors, like some captains, stay in the lanes, steering the production safely by rote and formula. Other directors, like the explorers James Cook and Louis-Antoine de Bougainville, sail off to discover something new and subsequently to change the charts. These directors, like Oliver Stone, Francis Ford Coppola, Martin Scorsese, Steven Spielberg, George Lucas, Francois Truffaut, Fredrico Fellini, and others, have the confidence to go where others have not, the stamina to complete the journey, and the artistry to expand the medium itself to accommodate their achievements.

But there is an important difference between a director of a motion picture and an ancient sea captain. Those captains that left the sea lanes could imagine where their journey might take them and what they might discover. But it did not matter what they envisioned. They would find whatever island, continent, or sea had been put there long before their arrival. However, a director who leaves the safe harbors must imagine the destination that lies beyond. If he does not, it will not exist. This destination is the only island or sea that the crew will ever find.

But imagining a destination is not enough to get there. In the same way a ship's captain would find it difficult to command a ship without knowing the names of the sails or what made a ship go forward, a director must know the technical aspects of film or video production to communicate with the crew.

The third element is the ability to lead. A director depends on the crew to make his ideas tangible. A good director has the confidence to inspire his crew to assist in achieving that goal. This is true also for a captain. The greatest sailor in the world will never reach his destination without inspiring his crew to follow orders. (Remember Captain William Bligh?)

It is clear that all three qualities are necessary in the art of directing. Film, like the other arts, begins with vision, and leadership and technical knowledge allow the artist to make that vision real. A great composer can never communicate his songs to others without being able to write music or to convince an orchestra to play his composition.

Is it possible to learn how to be a director, or are you "born" a director? From where does the music first begin to play? To help all of us "would-be" directors, Academy Award–winning director and screenwriter Oliver Stone helps us explore these questions.

Interview with Oliver Stone

Oliver Stone is an Academy Award–winning writer and director whose writing credits include Midnight Express, Scarface, *and* Year of the Dragon. *His writing/directing credits include* Salvador, Platoon, Wall Street, *and* Talk Radio. *Both critics and audiences acknowledge Mr. Stone to be one of the great writer/directors of our time.*

▼

Does the anger and frustration that comes from recognizing injustice provide the motivation and singlemindedness needed to succeed in becoming a motion picture director or writer in this highly competitive industry?

No. I am not up on a soapbox. I go with the people. The people come first. People, faces, that is what movies are about. Eisenstein showed it and Griffith. They dealt with faces. People are interested in people. Justice and injustice and issues and politics and ideologies are only of interest if the person who is interested in them is of interest.

What would you consider the best intellectual training for a director to be?

The best intellectual training for a director is world experience, life experience. Adversity, struggle, inner struggle, and outer struggle are all good. Education and reading help, but I do not think they are the primary tools.

In 1969, you went to New York University Film School on the GI Bill, and one of your teachers was Martin Scorsese. Is film school important for learning the technical aspects of production only, or is it important for learning about the narrative aspects as well? Would you consider it mandatory for all would-be directors to attend film school?

No, I do not think it is mandatory at all. You learn from struggle and from your own problems and from life. I think film school helped as a secondary tool. It helped me find the medium for ideas and shape them. It also provided a competitive climate where you can talk, meet, and share ideas with other people who are interested in the same thing. But as to narrative sense, I think that is something you teach yourself, or you have it, or you learn it by other ways.

   I remember a lot of the students [and] a lot of the older directors had a good technical sense but they did not have much of a writing or a story sense.

Where does good writing and story sense come from? What is the root of being a good storyteller? For example, can you say something about the process of turning a real-life character into a fictional one?

I never considered the goal as being of turning [a real-life character] into a fictional one. I always considered it taking a real character and making him more real. And trying to find those moments in his life that represent something significant and dramatic that you bring out to the audience. Taking something and making it clearer. Life is filled with twenty-four-hour obscurities. Taking moments, finding them and shining them up, polishing, and bringing them forth. Where that comes from, the writing sense, comes from being a writer. From actually having your ass in the seat and doing it. I am not saying all directors have to be writers. Some directors do have a good story sense, but they are able to supervise and work with the other writers.

Have you ever described a situation in which you have no personal knowledge or experience?

Several times that has happened to me. *Scarface* was wholly written (by me) but based on people whom I had only briefly met or knew. But it was certainly not a life experience for me.

Is the actual process of directing pleasurable?

Yes and no. I think the aspect that is pleasurable is bringing out something, educating somebody else, educating yourself. Finding a moment. Making something greater than it really was. Shining. I suppose the unpleasant aspects are sometimes the army of details to deal with. The sense of continual overwork and fatigue. All the money problems and the thousand and one choices you must make each day.

What advice would you have for film students who feel that they have something important to say and wish to use film to say it?

My advice is to keep working at it. Keep refining it. There is always another way of saying it. Not to be vain or conceited about it or to think that they have all the answers. The answers keep changing. It is the questions that have to be asked. The constant questioning. The constant attention to the inner struggle. The development as a human being as well as an artist. If denied the opportunity to film, then write. Writing comes cheaply. All you need is paper and a typewriter or a word processor. I stuck with it through many years when I always thought I should be a filmmaker but I was not able to, and I kept writing. I wrote a lot of scripts.

It takes a great deal of dedication and belief in what you are doing.

Nobody else believes in you at the beginning.

You mentioned that you are interested in people above and beyond a plot or a particular theme. To what extent should a director involve himself with the people who are involved in sociopolitical problems of the times?

I do not think there is any responsibility there. I think you have a responsibility to your pact, your destiny, and you have to march in that direction. You have to go where your drum takes you. Like Thoreau went where he had to go. That is the only way you are going to fulfill yourself as an artist. You have to keep growing, and sometimes you are going to grow in directions that other people may not want you to go.

So you write and direct for yourself and not for an audience?

That is correct. Ultimately, yes. You like to blend the two, but ultimately you have to make yourself happy.

Appendix A

Motion Picture/Television Industry Production Directories

This appendix lists production directories used in various areas of the United States. If a directory is not listed for your city or state, contact your local film liaison office (Appendix B) to find the production directory that covers your area.

California

Pacific Coast Studio Directory
6313 Yucca St.
Hollywood, CA   90028-5093
(213) 467-2920 $20/year (published quarterly)
Lists production companies and support services throughout the United States.

Brooks Commercial Production
   Directory
The Stanley J. Brooks Company
1416 Westwood Blvd., Suite 205
Los Angeles, CA   90024
(213) 470-2849 $29
Specializes in listings of support services throughout California.

LOS ANGELES

LA 411
P.O. Box 480495
Los Angeles, CA   90048
(213) 460-6304 $40
Specializes in production information for television and commercials.

The Studio Blu-Book
Hollywood Reporter
6715 Sunset Blvd.
Hollywood, CA   90028
(213) 464-7823 $40
Specializes in listings of production companies and some crew and location information.

SAN FRANCISCO

SF 411
P.O. Box 77146
San Francisco, CA 94107
(415) 285-5556 $40
Specializes in production information
  for television and commercials.

NORTHERN CALIFORNIA

The Reel Directory
P.O. Box 866
Cotati, CA 94928
(707) 795-9367 $15
Specializes in listings of crews and pro-
  duction companies.

Florida

Florida Production Guide
Florida Motion Picture and TV Office
101 E. Gaines, Fletcher Bldg., Room
  B72
Tallahassee, FL 32301
(904) 487-1100 $22
Lists production facilities throughout
  Florida.

Illinois

Illinois Production Guide
Illinois Film Office
100 Randolph St., Suite 3-400
Chicago, IL 60601
(312) 917-3600   Free
Specializes in crew listings.

Chicago Creative Directory
333 N. Michigan Ave., Suite 810
Chicago, IL 60601
(312) 236-7337 $40
Lists production companies, equipment
  sources, and support services.

Kentucky

Kentucky Film Office Production
  Directory
Kentucky Film Office
Berry Hill Mansion
Frankfort, KY   40601
(502) 564-3456   Free
Lists production companies, equipment
  sources, and support services.

Massachusetts

Massachusetts Production Guide
Massachusetts Film Bureau
Transportation Bldg., 10 Park Plaza,
  Suite 2310
Boston, MA   09116
(617) 973-8800   Free
Lists production companies, equipment
  sources, and support services.

Minnesota

Minnesota Film And Video Book
Minnesota Film Board
100 North 6th St., Suite 880C
Minneapolis, MN   55403
(612) 332-6493   Free
Lists production companies, equipment
  sources, and support services.

New York

UPSTATE

Ad Facs
P.O. Box 3933, Stuyvesant Plaza
Albany, NY   12203
(518) 283-3923 $9.95
Lists production companies, crews, and
    support services.

NEW YORK CITY

Madison Avenue Handbook
Peter Glenn Publications, Ltd.
17 East 48th St.
New York, NY   10017
(800) 223-1254 $40
Lists production companies throughout
    New York, the Midwest, the South-
    east, and other areas of the country.

New York Theatrical Sourcebook
Broadway Press
120 Duane St., #407
New York, NY   10007
(212) 693-0570 $25.50
Specializes in art department sources.

New York Production Guide (NYPG)
150 5th Ave., Suite 219
New York, NY   10011
(212) 243-0404 $45
Lists production sources and crew po-
    sitions throughout New York and the
    rest of the country.

Texas

Texas Production Manual
Texas Film and Music Office
P.O. Box 12728
Austin, TX   78711
(512) 469-9111 $27
Lists production companies, crew posi-
    tions, and support services.

United States

*See Pacific Coast Studio Directory (California), Madison Avenue Handbook
(New York), and The Producer's Masterguide (International).*

International

The Producer's Masterguide
New York Production Manual, Inc.
611 Broadway, Suite 807
New York, NY   10012
(212) 777-4002 $79.95
Lists production information for the
  United States and abroad.

Appendix B

Film Commissions and Motion Picture Liaison Offices

This appendix lists many of the film commissions in the United States. If a commission is not listed for your city or area, contact your local chamber of commerce to find the one nearest you.

Alabama

Alabama Film Commission
340 North Hull St.
Montgomery, AL   36130
(205) 261-4195

Alaska

Alaska Motion Picture & TV Production Services
3601 C St., Suite 722
Anchorage, AK   99503
(907) 563-2167

Arizona

Arizona Motion Picture Development
   Office
1700 W. Washington
Phoenix, AZ   85007
(602) 255-5011

Lake Havasu Film Commission
1930 Mesquite Ave., Suite 3
Lake Havasu, AZ   86403
(602) 453-3456

Phoenix Motion Picture Office
251 W. Washington
Phoenix, AZ   85003
(602) 262-4850

Scottsdale Film Commission
3939 Civic Center Plaza
Scottsdale, AZ   85251
(602) 944-2422

Tucson Film Office
P.O. Box 27210
Tucson, AZ   85726
(602) 791-4000

Arkansas

Arkansas Office of Motion Picture
   Development
1 State Capitol Mall
Little Rock, AR   72201
(501) 371-7676

California

California Film Commission
6922 Hollywood Blvd., Suite 600
Hollywood, CA   90028
(213) 736-2465

Big Bear Lake Film Commission
Chamber of Commerce
P.O. Box 2860
Big Bear Lake, CA   92315
(714) 866-6190

Brawley Film Commission
Chamber of Commerce
P.O. Box 218
Brawley, CA   92227
(619) 344-3160

Contra Costa County, Convention and
   Visitors Bureau
2151 Salvio St., Suite N
Concord, CA   94520
(415) 685-1184

Fresno County Film Commission
Economic Development Corporation
2310 Tulare St., Suite 235
Fresno, CA   93721
(209) 233-2564

Humboldt County Film Commission
Convention and Visitors Bureau
1034 Second St.
Eureka, CA   95501
(717) 443-5097

Kern County Film Commission
Kern County Board of Trade
P.O. Box 1312
Bakersfield, CA   93302
(805) 861-2367

Kings County Film Commission
Crown Development Corporation
1222 W. Lacey Blvd., Suite 101
Hanford, CA   93230
(209) 582-4326

Lake Tahoe (South), Lake Tahoe Visitors Authority
P.O. Box 16299
South Lake Tahoe, CA   95706
(916) 544-5050

City of Los Angeles Motion Picture Coordination Office
6922 Hollywood Blvd., Suite 600
Hollywood, CA   90028
(213) 485-5324

County of Los Angeles Motion Picture
Coordination Office
6922 Hollywood Blvd., Suite 600
Hollywood, CA   90028
(213) 738-3456

Madera County Film Commission
P.O. Box 126
Bass Lake, CA   93604
(209) 642-3111

Mammoth, Mono and Inyo County,
Mammoth Location Services
P.O. Box 24
Mammoth Lakes, CA   93546
(619) 934-2571

Mendocino County Development
Corporation
320 South State St.
Ukiah, CA   95482
(707) 463-0860

Merced Film Commission
P.O. Box 3107
Merced, CA   94344
(209) 384-3333

Modesto Film Commission
Chamber of Commerce
1114 H Street
P.O. Box 844
Modesto, CA   95354
(209) 577-5757

Monterey Peninsula Film Commission
County of Monterey
P.O. Box 180
Salinas, CA   93902
(408) 757-1533

Oakland Film Commission
Mayor's Film Liaison Office
505 14th St., Suite 644
Oakland, CA   94612
(415) 273-3109

Palm Springs Film Commission
City of Palm Springs
P.O. Box 1786
Palm Springs, CA   92263
(619) 323-8277

Pismo Beach Film Commission
Chamber of Commerce
581 Dolliver St.
Pismo Beach, CA   93449
(805) 773-4382

Placer County Film Commission
Business and Industry Development
  Commission
P.O. Box 749
Newcastle, CA 95658
(916) 663-2062

Redding Convention and Visitors
  Bureau
747 Auditorium Dr.
Redding, CA 96001
(916) 225-4100

San Bernadino County Film
  Commission
Board of Supervisors
385 North Arrowhead
San Bernadino, CA 92415
(717) 387-4828

San Diego Motion Picture and TV
  Bureau
110 West C. St., Suite 1600
San Diego, CA 92101
(619) 232-0124

Sonora Motion Picture Association
P.O. Box 382
Sonora, CA 95370
(209) 533-1651

Stockton Film Commission
Chamber of Commerce
445 West Weber
Stockton, CA 95203
(209) 466-7066

Colorado

Colorado Motion Picture Commission
1313 Sherman St., Room 523
Denver, CO 80203
(303) 866-2778

Boulder Film Commission
2440 Pearl St.
Boulder, CO 80302
(303) 442-1044

Colorado Springs Film Commission
P.O. Box 1575
Colorado Springs, CO 80901
(303) 578-6600

Connecticut

Connecticut Film Commission
210 Washington St.
Hartford, CT 06106
(203) 566-7947

Delaware

Delaware Development Office
99 Kings Highway
P.O. Box 1401
Dover, DE   19903
(302) 736-4254

Florida

Florida Motion Picture and TV Office
101 East Gaines
Fletcher Bldg., Room B72
Tallahassee, FL   32301
(904) 487-1100

Apalachicola Bay Chamber of
   Commerce
45 Market St.
Apalachicola, FL   32320
(904) 653-9419

Daytona Beach Area Chamber of
   Commerce
P.O. Box 2775
Daytona Beach Shores, FL   32015
(904) 255-0981

Broward Economic Development
   Board
One East Broward Blvd., Suite 1604
Fort Lauderdale, FL   33301
(305) 524-3113

Greater Key West Chamber of
   Commerce
P.O. Box 187
Lake Placid, FL   33852
(813) 465-6809

Beacon Council/Director of Film, TV
   and Radio
One World Trade Plaza
80 Southwest 8 St., Suite 2400
Miami, FL   33130
(305) 536-8000

Motion Picture and Television Indus-
   trial Development Commission of
   Mid-Florida Inc.
P.O. Box 2144
Orlando, FL   32802
(305)422-7159

Coordinator for Motion Picture and
   Television Development
City Hall Plaza, 8N
Tampa, FL   33602
(813) 223-8419

Georgia

Georgia Film Commission
230 Peachtree St.
Atlanta, GA   30303
(404) 656-3591

Hawaii

Hawaii Department of Planning and
  Economic Development
P.O. Box 2359
Honolulu, HW   96804
(808) 548-4535

County of Hawaii Department of Research and Development
34 Rainbow Dr.
Hilio, HW   96720
(808) 961-3366

County of Kuai
Office of the Mayor
4396 Rice St.
Lihue, HW   96766
(808) 245-3385

Counties of Maui, Molokai and Lanai
Maui Motion Picture/TV Coordinating
  Committee
P.O. Box 1738
Kahului, HW   96732
(808) 871-8691

Idaho

Idaho Film Bureau
Capitol Building, Room 108
Boise, ID   83720
(208) 334-4357

Illinois

Illinois Film Office
100 Randolph St., Suite 3-400
Chicago, IL   60601
(312) 917-3600

Chicago Film Commission
121 North LaSalle
City Hall, Room 812
Chicago, IL   60602
(312) 744-6415

Indiana

Indiana Department of Commerce
1 North Capitol St., Suite 700
Indianapolis, IN   46204
(317) 232-8821

Iowa

Iowa Development Commission
200 E. Grand Ave.
Des Moines, IO   50309
(515) 281-8319

Kansas

Kansas Film Commission
400 West 8 St., Suite 500
Topeka, KS   66603
(913) 296-2009

Kentucky

Kentucky Film Office
Berry Hill Mansion
Frankfort, KY   40601
(502) 564-3456

Louisville Film Office
Mayor's Office
601 W. Jefferson
Louisville, KY   40202
(502) 625-3061

Louisiana

Louisiana Film Industry Commission
P.O. Box 94361
Baton Rouge, LA   70804
(504) 342-8150

Kenner Film Commission
1801 Williams Blvd.
Kenner, LA   70062
(504) 468-7234

Thibidoaux Film Commission
P.O. Box 467
Thibidoaux, LA   70302
(504) 446-1187

Maine

Maine Film Commission
P.O. Box 8424
Portland, ME   04104
(207) 797-6991

Maryland

Maryland Film Commission
Department of Economic & Employment Development
45 Calvert St.
Annapolis, MD   21401
(301) 974-3577

Baltimore City Film Commission
303 East Fayette St., Suite 300
Baltimore, MD   21202
(301) 396-4550

Massachusetts

Massachusetts Film Bureau
Transportation Building
10 Park Plaza, Suite 2310
Boston, MA   09116
(617) 727-3330

Michigan

Michigan Department of Commerce/
   Film & TV
1200 Sixth St., 19th Floor
Detroit, MI   48226
(313) 256-9098

Minnesota

Minnesota Motion Picture/TV Film
  Board
100 North 6th St., Suite 880C
Minneapolis, MN   55415
(612) 332-6493

Minneapolis Office of Film, Video and
  Recording
323 M City Hall
Minneapolis, MN   55415
(612) 348-2491

Mississippi

Mississippi Film Commission
P.O. Box 849
Jackson, MS   39205
(601) 359-3037

Natchez Film Commission
311 Liberty Road
Natchez, MS   39120
(601) 446-6345

Columbus-Lownds County Film
  Commission
P.O. Box 789
Columbus, MS   39703
(601) 329-1191

Missouri

Missouri Film Commission
P.O. Box 118
Jefferson City, MO   65102
(314) 751-9050

St. Louis Film Partnership
100 South 4th St., Suite 500
St. Louis, MO   63102
(314) 444-1174

Montana

Montana Department of Commerce
1424 9th Ave.
Helena, MT   59620
(406) 444-2654

Lewistown Film Commission
211 West Main
Lewistown, MT   59457
(406) 538-9091

Butte Film Task Force
2950 Harrison
Butte, MT   59701
(406) 494-5595

Nebraska

Nebraska Telecommunications and Information Center
P.O. Box 95143
Lincoln, NE 68509
(402) 471-3368

Lincoln Film and Television Office
129 North 10th, Room 111
Lincoln, NE 68508
(402) 471-7375

Nevada

Nevada Motion Picture & TV Development
McCarran International Airport, Floor 2
Las Vegas, NV 89158
(702) 486-7150

Northern Nevada Picture and Television Division
600 East Williams St.
Carson City, NV 89710
(702) 885-4325

New Jersey

New Jersey Motion Picture and TV Commission
Gateway One, Suite 510
Newark, NJ 07102
(201) 648-6279

New Mexico

New Mexico Film Commission
1050 Old Pecos Trail
Santa Fe, NM 87501
(505) 545-9871

Albuquerque Film Commission
P.O. Box 1293
Albuquerque, NM 87103
(505) 768-4512

New York

New York Governor's Office for Motion Picture and TV Development
1515 Broadway, Floor 32
New York, NY 10036
(212) 309-0540

New York City Mayor's Office of Film
254 West 54th St.
New York, NY 10019
(212) 489-6710

Buffalo Film Commission
City Hall, Room 201
Buffalo, NY 14202
(716) 851-4841

Nassau County Film Commission
1550 Franklin Ave.
Mineola, NY 11501
(516) 535-4159

Suffolk County Film Commission
Dennison Bldg., Floor 11
Veterans Memorial Highway
Hauppauge, NY 11788
(516) 360-4800

North Carolina

North Carolina Film Office
430 North Salisbury St.
Raleigh, NC 27611
(919) 733-9900

North Dakota

North Dakota Economic and Development Commission
Liberty Memorial Bldg.
Bismarck, ND 58505
(701) 224-2810

Ohio

Ohio Film Bureau
P.O. Box 1001
Columbus, OH 43216
(800) 848-1300

Oklahoma

Oklahoma Film Office
Oklahoma Department of Commerce
6601 Broadway Extension
Oklahoma City, OK 73116
(405) 843-9770

Oregon

Oregon Film and Video Department
595 Cottage St., N.E.
Salem, OR 97310
(503) 373-1232

Portland Film Office
1220 S.W. 5th, Room 203
Portland, OR 97204
(503) 248-4739

Pennsylvania

Pennsylvania Film Bureau
449 Forum Bldg.
Harrisburg, PA 17120
(717) 787-5333

Philadelphia Film Commission
Municipal Services Bldg., Room 120
Philadelphia, PA 19102
(215) 686-2668

Rhode Island

Rhode Island Film Commission
150 Benefit St.
Providence, RI 02903
(401) 277-3456

South Carolina

South Carolina Film Office/State Development Board
P.O. Box 927
Columbia, SC 29202
(803) 737-1400

South Dakota

South Dakota Film Commission
Department of Tourism
Capitol Lake Plaza
Pierce, SD   57501
(605) 773-5032

Tennessee

Tennessee Film, Entertainment and
  Music Commission
The Rachel Jackson Bldg., Floor 7
Nashville, TN   37219
(615) 741-3456

Memphis/Shelby County Film, Tape and
  Music Commission
160 North Mid America Mall, Suite 660
Memphis, TN   38103
(901) 576-4284

Texas

Texas Film and Music Office
P.O. Box 12728
Austin, TX   78711
(512) 469-9111

El Paso Film Commission
1 Civic Center Plaza
El Paso, TX   79901
(800) 492-6001

Houston Film Commission
Greater Houston Convention and Visitors Bureau
3300 Main Street
Houston, TX   77002
(713) 523-5050

Film Commission of North Texas
6311 North O'Connor
Lock Box 57
Irving, TX   75039
(214) 869-7657

San Antonio Convention and Visitors
  Bureau
P.O. Box 2277
San Antonio, TX   78298
(800) 531-5700

Utah

Utah Film Development
6290 State Office Bldg.
Salt Lake City, UT   84114
(801) 538-3039

Southern Utah Film Commission
97 E. St. George Blvd.
Saint George, UT   84770
(801) 628-4171

Vermont

Vermont Film Bureau
134 State Street
Montpelier, VT   05602
(802) 828-3236

Virginia

Virginia Film Office
1000 Washington Blvd.
Richmond, VA   23219
(804) 786-3791

Washington

Washington Film and Video Office
312 1st Ave. N.
Seattle, WA   98109
(206) 464-7148

Washington, DC

Mayor's Office of Motion Picture &
   Television Development
1111 E Street N.W., #700
Washington, DC   20004
(202) 727-6600

West Virginia

West Virginia Governor's Office of Economic Development
2101 Washington St. E.
Capitol Complex, Bldg. 6
Charleston, WV   25303
(304) 348-2286

Wisconsin

Wisconsin Film Office
P.O. Box 7970
Madison, WI   53707
(608) 267-7176

Milwaukee Film Commission
809 North Broadway
Milwaukee, WI   53202
(414) 223-5818

Wyoming

Wyoming Film Office, I-25 at College
  Dr.
Cheyenne, WY   82002
(307) 777-7851

Appendix C

Motion Picture/Television Industry Trade Magazines and Newspapers

American Cinematographer
1782 N. Orange Dr.
Hollywood, CA   90028
(213) 876-5080
Covers the cinemagraphic arts.

Backstage
5150 Wilshire Blvd., Suite 302
Los Angeles, CA   90036
(213) 936-5200

Billboard
9107 Wilshire Blvd., #700
Beverly Hills, CA   90210
(213) 273-7040
Classifieds list new productions.

Casting Call
3365 Cahuenga Blvd.
Hollywood, CA   90068
(213) 874-4012
Lists new productions in the Los
   Angeles area.

The Daily Variety
1400 N. Cahuenga Blvd.
Hollywood, CA   90028
(213) 469-1141
Covers all "news" aspects of industry
   and new productions.

Drama Logue
P.O. Box 38771
Los Angeles, CA   90038
(213) 464-5079
Lists new productions casting and
   "crewing up" in the Los Angeles
   area.

Electronic Media Magazine
6404 Wilshire Blvd., #600
Los Angeles, CA   90048
(213) 651-3710

Hollywood Reporter
6715 W. Sunset Blvd.
Hollywood, CA   90028
(213) 464-7411
Covers all "news" aspects of industry
   and new productions.

In Motion Film & Video Magazine
421 4th St.
Annapolis, MD   21403
(301) 269-0605

Movie/TV Marketing
P.O. Box 7519
Northridge, CA   91327
(818) 368-0786

Motion Picture Almanac
159 W. 53rd St., #16E
New York, NY   10019
(212) 247-3100

Screen International
8500 Wilshire Blvd., #826
Beverly Hills, CA   90211
(213) 462-6775

Spotlight Casting
P.O. Box 3720
Los Angeles, CA   90078
(213) 462-6775
Lists productions that are casting and
  "crewing up" in the Los Angeles
  area.

Television Broadcast Magazine
P.O. Box 7926
Overland Park, KS   66207
(913) 642-6611

Variety
154 W. 46th St.
New York, NY   10036
(212) 869-5700

Glossary

This glossary lists many of the words, terms, and titles used frequently in motion picture/television production. This list is not comprehensive but was compiled by selecting those words heard commonly during production. For a more comprehensive list of film and video terminology, there are several excellent production dictionaries available.

Above the line   A production budget is divided into above the line and below the line. Above the line usually contains the cost of the "creative" elements, such as the script, the director, the actors, and the producer's fees. (*See* below the line.)

A.C.E.   American Cinema Editors. An honorary society for professional film editors.

Action   A command given by the director when she wants the movement and dialogue to begin in a scene.

A.D.   Assistant director. (*See* first assistant director, second assistant director.)

Aerial shot   A scene filmed from the air (i.e., from a plane or helicopter).

AFI   American Film Institute.

Air to air   Filming or taping a flying object (e.g., plane) from another flying object (e.g., helicopter).

Apple box   A wood crate used to raise props, people, lights, etc., during shooting. There are also half apples and quarter apples.

Art Department   The department that is responsible for creating the look of the film or video through sets and props.

A.S.C.   American Society of Cinematographers.

Assistant cameraman   *See* first assistant cameraman, second assistant cameraman.

Assistant director *See* first assistant director, second assistant director.

Atmosphere Extras used in a scene.

Auditor The production accountant or controller. Responsible for the production's bookkeeping and estimating.

Baby legs A short tripod used to shoot from low angles.

Barn doors Folding metal gates attached to a light, used to direct the light and control the intensity.

Below the line Production costs for crews and materials, stock, processing, and postproduction expenses.

Best boy Best boy electrical is second in command of the electrical department (gaffer is first) and is responsible for the other electricians and electrical equipment. Best boy grip is second in command of grip department (key grip is first) and is responsible for the grip equipment and other grips.

Blacks Cloths hung in windows and doorways used to block out daylight to create the illusion of night.

Blocking When a director practices the action in a scene with the actors or the camera.

Boards *See* production boards.

Body makeup artist According to union guidelines, a union makeup artist can only apply makeup to an actor's face, head, forearms, and hands. All other areas of the body are the responsibility of the body makeup artist.

Boom A type of extension pole used to hold a microphone.

Breakaway Props used in stunts that are designed to break apart on impact.

Breakdown Reading the script and looking for information relevant to the responsibilities of your position. (For example, the location manager reads the script for the type and number of locations suggested.)

Butterfly *See* silk.

Cable Heavy-duty insulated wires used to conduct electricity from a power source (e.g., generator) to equipment (e.g., lights). It usually comes in 25-, 50-, and 100-foot lengths.

Cable run The distance or path of the cables from the power source to the set.

Call   The location and time of the next day's (or night's) filming or videotaping.

Call sheet   Sheet that is usually handed out to cast and crew members to announce the next day's shooting schedule, each cast and crew member's call time (what time to report to the set), which sets will be used, and any unusual equipment required.

Camera car   A special vehicle on which a camera can be mounted and used to film or videotape moving vehicles or persons.

Cameraman   *See* director of photography.

Camera mount   A device used to mount the camera to a tripod, dolly, or crane.

Camera operator   Also called the second cameraman. Responsible for running the camera and keeping the action in frame.

Camera tracks   Also called dolly tracks. The metal or wooden rails on which a dolly rides.

Casting director   The person who specializes in finding the best actor for each speaking role in the production and negotiating their contracts. Choices depend on approval by the director and producer.

Changing bag   A lightproof bag used by the film loader or second assistant cameraman to load film into the magazine in daylight.

Cinematographer   *See* director of photography.

Clap sticks   The part of the slate or clapper board that is clapped down to make a sound. Used to provide a sync cue, connecting picture with sound. (*See also* slate.)

Closed set   1. A set that is closed to the public. 2. A set in which an intimate scene is shot (i.e., nudity) that is closed to most of the crew during the actual shooting.

Closeup   A shot that is framed to show the details of a subject.

Cone lights   Large, cone-shaped flood lights that illuminate a large area with soft light.

Construction crew   Crew composed of prop makers and laborers who are responsible for constructing sets. Supervised by a construction foreman.

Contact list   1. Production contact list that lists names, telephone numbers, and addresses of all supply houses and services used by the production (e.g., a film laboratory). 2. Location contact list is a compilation of every practical location with the corresponding address, contact person (i.e. owner), and telephone number.

Continuity   Responsibility of script supervisor; keeping track of the picture and sound details so that the story is developed consistently. (*See* script supervisor.)

Costume designer   The individual who creates the costumes worn by the cast on a production.

Cover set   An alternate set (usually interior) that can be used when bad weather prevents shooting.

Craft service   Department that is responsible for providing the crew with beverages and snacks (e.g., coffee, doughnuts, apples) during shooting.

Crane   A piece of equipment used to carry the camera and two people or more high into the air for high-angle shots.

Cut   Said by the director to stop the action during filming or taping.

Day for night   Shooting exterior night scenes during the daylight hours by using special blue filters.

Deal memo   A binding written statement that outlines the basic terms of an agreement (e.g., salary, time schedule, screen credit).

Deuce   A 2000-watt spotlight.

DGA   Director's Guild of America.

Diffused light   Soft, shadowless light created by placing diffusers in front of a light source.

Dinky-inky   A small spotlight, usually 100 to 200 watts.

Director   The person with ultimate authority over all creative aspects of a production.

Director of photography   Also called D.P., cinematographer or cameraman. Responsible for all the photographic aspects of the production, including lighting the scene and composing the shot (in collaboration with the director).

Director's Guild of America   Union for directors, assistant directors, and production managers.

Dolly   A mobile platform with wheels for use in moving the camera and its operator.

Dolly shot   A shot where the camera moves on a dolly while filming or videotaping.

Double   A person who replaces an actor to perform nonacting scenes. (*See* stunt double.)

Down time   Time lost during shooting when equipment must be changed or repaired or the company must move to another location.

Drive on   When a studio employee leaves permission with the gate guard to allow an individual to drive onto the studio lot.

Dulling spray   A spray used to give reflective surfaces a matte finish so that they will not reflect light into the camera and create a hot spot or flare.

Duvatyne   Black cloth used in devices (i.e., flags) to create shadows.

Electrician   Member of the electrical department under the supervision of the gaffer. Responsible for powering, setting, and adjusting the lights.

Establishing shot   A shot used in the beginning of a sequence to specify the location by providing a "full" view of the scene.

Executive producer   A title that can be given to any number of people associated with the picture, including the financier, the U.P.M., or a friend of the producer.

Expendables   Items purchased for use on a film that will probably be consumed or used up over the course of the production (e.g., tape, gels, wire).

Exterior   Abbreviated EXT. Any scene filmed outside.

Extra   A nonspeaking part (unless spoken with a group) used to fill in and add realism to a scene.

Feature length   A full-length film or video usually over 80 minutes long (i.e., a movie).

Field of view   The area seen by the camera.

Fill light   Secondary lights used to fill in the shadows created by other light sources.

First assistant cameraman   Member of camera crew whose responsibilities include changing lenses, maintaining focus during shooting, and measuring the distance between the lens and the objects being photographed.

First assistant director   Responsible for making sure that everything needed for shooting is at the right place at the right time. During preproduction the first A.D., with the director's and production manager's approval, creates the production schedule or production boards, hires extras, and schedules technical scouts. During the shoot, the first A.D. is the liaison between the director and the rest of the crew and is responsible for the call and for maintaining discipline on the set. The first A.D. yells, "Quiet on the set" and tells the camera operator to "roll."

Flag   A sheet of duvatyne set in a frame and used to shade an area of the set.

Flare   Bright spot on exposed film or videotape usually caused by lights reflecting off a shiny surface.

Flat   A movable section of a set (e.g., a fake wall).

Flat rate   A complete and total fee paid for services rendered that will not provide any additional amount for overtime or weekend work.

Focus puller   A member of the camera crew (usually the first assistant cameraman) who is responsible for maintaining focus while shooting.

Fog machine   A device that creates fog.

Foley   A sound or sound effect that is recorded in a studio after shooting and is edited into the sound track (e.g., knock on the door, a ringing telephone).

Foley artist   A specialist who can recreate certain sounds to be used in the sound track of a film or video.

Foot   The end of a reel of tape or film.

Forced call   A violation of union contract by establishing a crew call that does not give the crew members a minimum number of hours for rest.

Freelance   To be hired on a per-job basis instead of being on staff or under long-term contract to a company.

Friction head   A device attached to a tripod to provide smooth camera movement while panning (horizontal movement) or tilting (vertical movement).

Front car mount   A device used to attach a camera to the hood of a car. A scene can be filmed in this manner while the car is moving.

Full shot   A shot that frames the subject from head to toe.

Gag   Stunt.

Gaffer   The chief electrician, who works under the director of photography and is responsible for lighting the set.

Gaffer's tape   A durable silver cloth tape that will stick securely to most surfaces. Also called electrician's tape or duct tape.

Gel   A diffusing material or color transparency used to soften or change the color of a light source.

Generator   A mobile source of electricity usually employed when shooting on location. Also called a "jenny."

Golden time   Overtime payment that can be twice or more the normal hourly rate.

Greensman Member of the crew who is responsible for dressing the set with plants, trees, bushes, etc., and maintaining them (e.g., watering).

Grip 1. Department responsible for moving camera, creating shadows, and rigging. 2. General term for crew members who provide labor on the set (i.e., stagehands).

Hairdresser Also called hairstylist. Individual who is trained and licensed to cut and color actors' hair to create or enhance their "look" in a production.

Hair in the gate Foreign matter, dust, dirt, etc., in the camera gate.

Half apple An apple box only half as tall as a full or normal apple.

Hand props Portable items such as a briefcase or newspaper carried by an actor in a scene and provided by the prop master or prop department. (*See also* prop.)

Hazard pay Increase in daily rate or additional payment when a crew or cast member is asked to work under conditions that are considered dangerous (e.g., underwater, special stunts, aerial).

HMI Light Halogen medium iodide. A portable, high-intensity lamp that emits daylight balanced light.

Honeywagon A large vehicle similar to a motorhome that has dressing rooms and toilet facilities.

Hot set A set that is being used for shooting or is fully prepared for shooting (i.e., completely dressed).

IATSE International Alliance of Theatrical and Stage Employees (abbreviated IA). The parent union that encompasses all below the line, distribution, and exhibition job categories. There are nearly 1000 IA locals throughout the United States.

Independent contractor A freelancer who works for a company and is paid the full amount for services rendered. (The paycheck contains no tax deductions. Independent contractors are responsible for withholding their own taxes.)

Independent producer 1. A producer who uses a production crew that is not under contract to a major studio. 2. A producer not under contract to a major studio. 3. A producer who finances a film without using the funding of a major studio.

Independent production A production not financed by a major studio or using crew members under contract to the studio.

Indie Also indy. Slang for independent producer or independent production.

Insert A closeup shot that is used to enhance continuity or explain part of the action that is later edited into a sequence (e.g., a vase crashing to the floor, closeup of a single tire skidding on the road).

Int Interior.

In the can 1. Exposed film. 2. A scene in which photography has been completed. 3. A film or video that has completed principal photography.

Jenny *See* generator.

Juicer Electrician.

Keg A 750-watt spotlight.

Key grip Head of the grip department. The key grip works closely with the director of photography and the gaffer.

Kook A screen placed in front of a light source and used to cast a variety of shadows. (Also cookie.)

Laboratory Where exposed film is processed and printed. (Also lab.)

Lead Cast members who are featured or "starring" in a production.

Lead man Head of the swing gang in the set dressing department.

Line producer "Hands on" producer involved in physical production of film or video. Oversees above the line and below the line personnel and elements.

Location A locale outside a studio backlot and stage that is used for filming or videotaping.

Location manager Crew member responsible for finding and securing all practical filming locations, negotiating rental fees, obtaining government film permits, and organizing all details related to the location (e.g., parking, police, and fire safety officers).

Magazine A film container or cassette that mounts on the camera and contains compartments for unexposed and exposed film. The magazine is loaded with film in a darkroom or lightproof changing bag. (Also "mag.")

Magic hour Twilight.

Makeup artist Crew member responsible for applying makeup to the cast.

Matte A technique that keeps specific areas of a film unexposed so that those areas can be later combined with other images.

Mixer Chief sound recordist on the set, who is responsible for recording the best quality sounds possible.

Monitor A video screen used to watch the images being taped.

Mos Filming or videotaping without recording synchronized sound.

NABET National Alliance of Broadcast Engineers and Technicians. National labor organization that represents technicians, artists, and engineers for radio, television, and made-for-television films. Like IATSE, NABET is affiliated with the AFL-CIO.

NG No good. What the director might say when he does not like the take and does not want it printed.

Night for night When sequences in a script that are supposed to take place at night are actually filmed at night (e.g., large night exterior scenes).

Nontheatrical Films or videos that will not be distributed to movie theatres but instead will undergo a limited distribution to the various nontheatrical markets (e.g., television, videocassette, cable television).

Off camera What is not seen by the camera.

Off screen A sound or action that takes place off camera so it will not be seen.

On call A crew or cast member who must be available to work if needed.

P.A. *See* production assistant.

Pan Movement of camera along the horizontal axle. (*See* tilt.)

Parallel A platform used to raise camera and crew above ground level to achieve a high shot.

Per diem An amount of money paid to production members based on the costs incurred by members of the crew and cast when shooting on location (e.g., cost of hotel and food).

Pickup 1. To reshoot a scene by not returning to the very beginning of the take. 2. Scene shot after principal photography has been completed.

Playback Previously recorded music that is played back (usually in a tape recorder) when shooting a musical sequence.

Postproduction   The period after principal photography when everything needed to present a finished and completed production is assembled (e.g., editing, looping, music, titles).

Practical   1. A prop that actually works. 2. A light on the set that works (e.g., a coffee table lamp).

Practical location   A real location used for filming that is outside a stage or studio backlot (e.g., a house, restaurant, school).

Preproduction   The period before principal photography when everything is prepared for shooting (e.g., designing costumes, casting actors, finding locations, designing sets).

Principal photography   The period when all material involving the major speaking parts or major scripted elements is filmed or taped.

Producer   The individual who "spearheads" a production and may be involved in most or all of the following activities: arranging for financing, hiring the writer, hiring the director, hiring the actors, overseeing the production, and supervising the release or distribution.

Production accountant   A staff or managerial position whose responsibilities include maintaining the production's financial records, estimating the production's cost, and supervising the crew's financial responsibilities. (Also production auditor, controller.)

Production assistant   Crew position in which the individual supports and assists many of the departments on a production. (Also P.A.)

Production boards   A master schedule created and used by the first A.D. or production manager. The "boards" are composed of thin strips of paper that represent each scene in the production. The strips are arranged on the board to provide the most efficient and least expensive shooting schedule and provide production information, including the duration of the shoot and the order and dates of the scenes to be shot.

Production coordinator   Individual who is responsible for acting as a clerical liaison between the production manager and the other departments during production. (Also production office coordinator.)

Production designer   Individual who supervises the art department and is responsible for the production's overall "look."

Production manager   Individual in charge of managing the entire production, overseeing all departments, and supervising all administrative and financial details of the production and the crew. (Also unit production manager [U.P.M.].)

Prop   1. Any portable or handable item used in a scene. 2. Any item specifically mentioned in the script.

Property master   Individual responsible for procuring and placing all props on the set.

Prop maker   Individual in the construction department responsible for constructing the props needed for a production.

Quarter apple   An apple box one fourth as high as a standard apple box.

Rack focus   Changing the focus during shooting to maintain focus on a particular subject.

Rails   1. Dolly tracks. 2. Scaffolds used on the set to raise lights.

Reflector   1. A silver or gold reflective panel used to reflect light from a light source onto a subject. 2. White cards used to direct light. (Also called bounce boards.)

Runners   1. Scaffolding on which lights, curtains, and other equipment can be hung. 2. Production assistants.

Run through   Complete rehearsal. (Also walk through.)

SAG   *See* Screen Actors Guild.

Sandbag   Divided bag filled with sand, used to weigh and hold down pieces of equipment.

Scale   Minimum rate of pay for a particular job as established by the union.

Screen Actor's Guild   The actor's union. SAG sets rates of pay and other standards including benefits.

Screen Extras Guild   The union for extras. (Also SEG.)

Scrim   Translucent screen made of wire gauze and placed in front of a light source to diffuse its intensity.

Script supervisor   Individual responsible for keeping detailed notes regarding each take, including dialogue, action, makeup, and lens used. These notes are extremely important to maintain the production's continuity during shooting and editing. (Also called continuity supervisor.)

Second assistant cameraman   Member of camera department who is responsible for loading and unloading magazines from the camera, filling out camera reports, preparing the slate for each take, and assisting the first assistant cameraman.

Second assistant director Assists the first assistant director; responsibilities include preparing and distributing call sheets and other production paperwork and placing extras on the set.

Second unit 1. Additional production crew that shoots scenes that do not use any actors in any of the leading roles (e.g., inserts). 2. When multiple cameras are used.

SEG *See* Screen Extras Guild.

Set Where a film or tape is being shot.

Set decorator Individual responsible for dressing the set (e.g., furniture, decor).

Set dressing Furniture, decor, etc., that is used to "dress" or decorate a set.

Setup The correct arranging and positions of all elements on a set before shooting (i.e., actor, lights, props, dressing).

Shoot Filming or videotaping. (Also shooting.)

Shooting schedule A schedule that lists the location, date, actors, and equipment needed to shoot the film or video. Sometimes the shooting schedule is created from the production boards. (*See* production board.)

Short subject A film or video that usually has less than thirty minutes of running time.

Shotgun mic A directional microphone that is sensitive to sound originating from a specific location.

Side car mount A piece of equipment that allows a camera to attach onto the side of a vehicle.

Signatory A production company that must obey the rules and obligations of a union or guild because it has signed an agreement with the union or guild to employ its members.

Silk A material used to diffuse bright sunlight.

Slate Board that may be photographed before or after each take. On the board is written the director's and cinematographer's name and the roll, scene, and take number. The board is then clapped together to provide a synchronous cue connecting the sound and picture. (Also clapperboard.)

Sound stage A soundproof room or building where scenes may be filmed or taped.

Special effects Any effect in a production that must be specially created (e.g., interplanetary voyages, explosions, fires).

Speed   1. What the mixer says to the director when the sound recorder is running at the proper speed to record sound that will be synchronous with the picture.

Spider box   Electrical junction box that provides several outlets. Usually used to power lamps.

Squib   A special electrical device that causes an object or person to appear as if hit by bullets. Sometimes a small packet of fake blood is built into the squib, so that when it is ignited the blood will come out.

Stand-in   An individual who takes the place of the actor to help the technicians properly set up and light the scene.

Steadi-cam   Balanced body frame that can be used with a hand-held camera. The frame allows an operator to follow the action and maintain a steady, fluid framing.

Still photographer   Individual responsible for shooting photographic stills (usually 35 mm) during production. These stills can be used for publicity or for helping to match continuity.

Storyboard   A series of drawings (similar to a comic strip) that represent each shot in a production. Storyboards are helpful in assisting the director to create a rhythmic progression of shots, communicate them to the crew, and ensure that an important shot is not forgotten.

Strike   To take apart a set after filming is over and return the location to its original condition.

Stunt coordinator   Individual responsible for coordination, supervision, and safe execution of all stunts on a production.

Stunt double   A stunt person who is made to look like an actor and takes the place of the actor when potentially dangerous scenes are shot. (Also stuntman, stuntwoman.)

Taft-Hartley law   A law that allows anyone to work in a union production for thirty days before being required to join that union.

Take   One segment of photography (e.g., from starting the shot to ending the shot or from "action" to "cut") during filming or taping.

Talent   All performers on a production (actors, extras, animals).

Teleprompter   An electronic cuing machine that flashes the scripted words on a monitor near the camera or on a mirror that fits over the camera lens. It is used in place of cue cards.

Tilt   Movement of the camera on the vertical axis.

Tracking shot   When the camera is moved to follow the action (e.g., the camera is mounted on a dolly and rolled or is mounted on a moving vehicle). (Also traveling shot.)

Transportation   Department in production responsible for all equipment vehicles needed behind the scenes (e.g., motorhomes, equipment trucks), all vehicles used in front of the cameras (e.g., picture cars), and for transporting the cast and crew when necessary.

Turnaround time   According to a union contract, the minimum amount of time off between ending work and beginning work. If this minimum amount of time is not provided, it becomes a forced call and the production will incur a financial penalty.

Two-shot   A shot that is just wide enough to frame two people simultaneously.

Union   A labor organization that sets standard minimum rates for pay, work conditions, benefits, and so on.

U.P.M.   *See* production manager.

Utility   A member of the crew who assists many of the other departments during production. (*See also* production assistant.)

Video assist   A video recording system used in conjunction with filming to view scenes immediately after they are shot.

Wigwag   A light outside a soundstage. When the light is turned on, it indicates that shooting is in progress and that there should be no interruptions. (Also warning light.)

Wild picture   A visual portion of a film or video shot without synchronized sound.

Wild track   Sound that is recorded without synchronized picture (e.g., room tone, off-camera dialogue). (Also wild sound.)

Working title   The tentative and temporary title given to a production before its final name is chosen and the production is released.

Wrap   1. When shooting is finished at the end of the day. 2. Completion of principal photography.

Bibliography

Careers in Film and Video Production *is designed as a pragmatic information source for those individuals who desire to better understand the inner workings of production and wish to develop a career in the motion picture/television industry. Ultimately, the most practical knowledge of film and video production comes from working on the set. However, many times books can enhance and supplement what is learned on the job. In addition, books are the best source of information for those areas in which you may not have any direct contact. The books listed below will broaden your understanding of the process by which films and videos are made.*

PREPRODUCTION

Baumgarten, Paul A., and Farber, Donald C. *Producing, Financing and Distributing Film,* 2nd ed. New York: Limelight Editions, 1987.

Bronfeld, Stewart. *How to Produce a Film.* Englewood Cliffs, NJ: Prentice-Hall, 1984.

Chamness, Danford. *The Hollywood Guide to Film Budgeting and Script Breakdown,* 4th ed. S.J. Brooks, 1981.

Garvy, Helen. *Before You Shoot: A Guide To Low-Budget Film Production.* Shire Press, 1988.

Gregory, Mollie. *Making Films Your Business.* New York: Schocken, 1979.

Lazarus, Paul. *The Movie Producer: Handbook for Producing and Picture Making.* New York: Harper & Row, 1985.

Robertson, Joseph F. *Motion Picture Distribution Handbook.* Blue Ridge Summit, PA: TAB Books, 1981.

Singleton, Ralph. *Film Scheduling.* Beverly Hills, CA: Lone Eagle Publications, 1984.

Squire, Jason E., ed. *The Movie Business Book.* Englewood Cliffs, NJ: Prentice-Hall, 1983.

Wiese, Michael. *Film and Video Budgets.* Boston: Michael Wiese Film Productions (Focal Press), 1984.

Wiese, Michael. *Home Video: Producing for the Home Market.* Boston: Michael Wiese Film Productions (Focal Press), 1986.

PRODUCTION

Alkin, Glyn. *Sound Recording and Reproduction.* Boston: Focal Press, 1981.

Alten, Stanley. *Audio in Media,* 2nd ed. Belmont, CA: Wadsworth, 1986.

Armer, Alan A. *Directing Television and Film.* Belmont, CA: Wadsworth, 1986.

Bare, Richard L. *Film Director: A Practical Guide to Motion Picture and Television Techniques.* New York: Macmillan, 1973.

Bernstein, Steven. *The Technique of Film Production.* Boston: Focal Press, 1989.

Burrows, Tom, and Wood, Donald. *Television Production—Disciplines and Techniques.* Dubuque, IO: William C. Brown, 1978.

Carlson, Verne, and Carlson, Sylvia. *Professional Cameraman's Handbook,* 3rd ed. Boston: Focal Press, 1981.

Carlson, Verne, and Carlson, Sylvia. *Professional Lighting Handbook.* Boston: Focal Press, 1985.

Carson, Richard. *Stage Make-up,* 7th ed. Englewood Cliffs, NJ: Prentice-Hall, 1986.

Detmers, Fred H., ed. *American Cinematographer Manual,* 6th ed. Hollywood, CA: ASC Press, 1986.

Dmytryk, Edward. *On Screen Directing.* Boston: Focal Press, 1983.

Eargle, John. *The Microphone Handbook.* Elar Publishing Company, 1982.

Fuller, Barry, Steve Kanaba, and Janyce Brisch-Kanaba. *Single-Camera Video Production Handbook.* Englewood Cliffs, NJ: Prentice-Hall, 1982.

Gradus, Ben. *Directing—The Television Commercial.* Boston: Focal Press, 1981.

Kehoe, Vincent J-R. *The Technique of the Professional Make-up Artist.* Boston: Focal Press, 1985.

Lukas, Christopher. *Directing for Film and Television.* New York: Doubleday, 1985.

Lipton, Lenny. *Independent Filmmaking.* New York: Simon & Schuster, 1983.

Malkiewicz, J. Kris. *Cinematography,* 2nd ed. Englewood Cliffs, NJ: Prentice-Hall, 1988.

Malkiewicz, J. Kris. *Film Lighting: Hollywood's Leading Cinematographers Talk About Their Work.* Englewood Cliffs, NJ: Prentice-Hall, 1986.

Miller, Pat. *Script Supervising and Film Continuity.* Boston: Focal Press, 1986.

Millerson, Gerald. *Video Production Handbook.* Boston: Focal Press, 1987.

Millerson, Gerald. *The Technique of Lighting for Television and Motion Pictures.* Boston: Focal Press, 1985.

Millerson, Gerald. *TV Scenic Design.* Boston: Focal Press, 1989.

Pincus, Edward, and Ascher, Steven. *The Filmmaker's Handbook.* New York: New American Library, 1984.

Rabiger, Michael. *Directing: Film Technique and Aesthetics.* Boston: Focal Press, 1989.

Roberts, Kenneth H., and Sharples, Win, Jr. *A Primer for Filmmaking.* Pegasus, 1971.

Rowlands, Avril. *The Production Assistant in TV and Video.* Boston: Focal Press, 1987.

Rowlands, Avril. *Script Continuity and the Production Secretary.* Boston: Focal Press, 1977.

Samuelson, David. *Motion Picture Camera and Lighting Equipment,* 2nd ed. Boston: Focal Press, 1986.

Samuelson, David. *Motion Picture Camera Techniques,* 2nd ed. Boston: Focal Press, 1984.

Silver, A., and Ward, E. *A Film Director's Team.* New York: Arco Publishing, 1983.

Ulmer, Shirley, and Servilla, C.R. *The Role of the Script Supervisor in Film and TV.* New York: Hastings House, 1986.

White, Hooper. *How to Produce Effective TV Commercials,* 2nd ed. Lincolnwood, IL: National Textbook Company, 1986.

Wiese, Michael. *The Independent Film and Videomaker's Guide.* Boston: Michael Wiese Film Productions (Focal Press), 1986.

Wurtzel, Alan. *Television Production,* 2nd ed. New York: McGraw-Hill, 1983.

Zettl, Herbert. *Television Production Handbook,* 4th ed. Belmont, CA: Wadsworth, 1984.

POSTPRODUCTION

Anderson, Gary. *Video Editing and Post-Production: A Professional Guide,* 2nd ed. White Plains, NY: Knowledge Industry Publications, 1988.

Balmuth, Bernard. *Introduction to Film Editing.* Boston: Focal Press, 1989.

Browne, Steve. *Videotape Editing: A Post-Production Primer.* Boston: Focal Press, 1989.

Dmytryk, Edward. *On Film Editing.* Boston: Focal Press, 1984.

Kerner, Marvin. *The Art of the Sound Effects Editor.* Boston: Focal Press, 1989.

Lustig, Milton. *Music Editing for Motion Pictures.* New York: Hastings House, 1980.

Reisz, Karl, and Millar, Gavin. *The Technique of Film Editing,* 2nd ed. Boston: Focal Press, 1968.

Robertson, Joseph F. *The Magic of Film Editing.* Blue Ridge Summit, PA: TAB Books, 1983.

Rosenblum, Ralph, and Kuren, Robert. *When the Shooting Stops . . . the Cutting Begins: A Film Editor's Story.* New York: Da Capo, 1986.

Schneider, Arthur. *Electronic Post-Production and Videotape Editing.* Boston: Focal Press, 1989.

Index

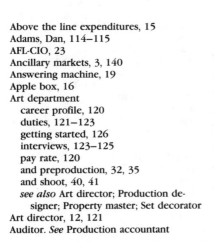